ANUNNAKI THEORY AUTHENTICATED WITH A NEW TWIST

Historical Evidence of Anunnaki Presence

Author

Shafak GokTurk

Translated by

Elizabet Narin Kurumlu

Publisher

Cosmo Publishing

All rights reserved, including the rights of reproduction in whole or in part of any form

ISBN: 978-1720942207

This book is dedicated to my father.

PREFACE 1

SECTION 1

WHICH ONE IS THE SIMULATION: THE HUMAN BEING OR THE UNIVERSE? 9

    The Limitations of Junk DNA and Simulation 10

    History of Homo Sapiens Sapiens 13

    Micro and Nano Robots 17

    One Hundred Spectacular Days God Granted 21

    The Great Code in the Junk Genes 23

    Sources: 26

SECTION 2

WHO ARE THESE ANUNNAKI? 29

    Ancient Ancestors; The Anunnaki 30

    Are the Anunnaki a Race from Space? 34

    Face-to-Face Contact with the Anunnaki 41

    About the Discovery of Nibiru/ Planet X, the Planet of the Anunnaki 44

    The Book of Genesis is being written 46

    The Deities of the Polytheistic Times: The Anunnaki 49

    The Anunnaki and the Rank-Frequency Distribution 51

    Sources: 53

SECTION 3

THE ROYAL FAMILY OF THE ANUNNAKI 55

    The Hierarchial Order with the Anunnaki 56

    The Nibirun King: An/Anu 58

    King of the Earth: Enlil 60

    The Mother of Life: Ninmah 62

    The Leader of the Age of the Aquarius: Enki 63

    A Healer and a Physician: Bau 65

    Sources: 66

SECTION 4

THE ANUNNAKI BASES IN THE SKIES 67

    Is the secret to longevity on Mars? 68

    What's happening on the Moon? 76

    The Traces of the Anunnaki on the Moon 81

    Did the Deluge really happen? 84

    The stepping stone of Humanity: Mars 88

    Is Mars going through Ice Age? 92

    The Monolith of Phobos and the Human Beings meeting with the Ancestors: The Accident of Phobos 95

    Sources: 100

SECTION 5

ENKI - ADMIRAL BYRD MEETING 103

    Was it Enki who met with Admiral Byrd? 104

    Admiral Richard B. Byrd's Diary, February - March 1947 105

    Flight Log – Base Camp Arctic – 2/19/1947 106

    Sources: 123

SECTION 6

MESSAGES FOR TODAY FROM THE BOOK OF ENOCH 125

    The Book of Enoch: 2000s 126

    The Authenticity of the Book of Enoch 128

    Enoch Says: This wisdom teaching is an appraisal of Earth for the next generations 130

    Wisdom, Light, Joy, Peace 131

Enoch in Enlil's Spacecraft 132

A volcano in the Mediterranean 137

The Hollow Earth Theory and the Country in the North 137

Enoch goes to the Garden of Aden over the Sea of Eritrea ... 138

Enoch in Machu Picchu 138

The Gates of Heavens; Is it the secret of going into space? 140

The holy race to descend from the Heavens pretty soon: The Anunnaki 142

Enoch in the mineral depots of the Anunnaki 143

Enlil's confession 144

Those Who Do Not Sleep in Heavens 146

Ancestors pre-dating History; Alexander and Enoch 147

The Chariots in the Skies: UFOs 152

Sources: 160

SECTION 7

SHAMS, MEVLANA AND THE ANUNNAKI 161

The Secret of Shams 162

The Meeting of Mevlana and Shams 166

The mysterious relationship between Shams and Mevlana .. 169

The envied bonding between Mevlana and Shams 172

The mysterious visitors of Mevlana and Shams 172

The sand of Hejaz on Mevlana's feet 176

The departure of Shams 177

Did Shams die? Or did he get slayed? Or did he leave? 179

Makalat 180

If Shams is an Anunnaki, why might he have visited Mevlana? 181

Sources: 184

SECTION 8

THE PYRAMID WAR AND THE ILLUMINATI(8760 BCE) 185

The three pyramids and the Sphinx in Giza: Who do they belong to? 187

Right. What about the Sphinx? 188

The pyramids of the Deities and their copies 190

The Great Pyramid War 194

Marduk is resisting in the Great Pyramid 197

The one who enters the Great Pyramid last is Ninurta 203

Sources: 206

SECTION 9

AN ALTERNATIVE HISTORY OF SIX THOUSAND YEARS 207

From the War at the Great Pyramid to the Illuminati 208

The Sumer Civilization 212

Sumer Civilization being established by Enlil 215

Egyptian Civilization being established by Enki (Ptah) 216

The Council of Twelve makes an Agreement based on the Constellation 218

Marduk's leadership is accepted in the Age of Ram 219

Towards the Age of Pisces 223

The Age of Pisces begins 225

Towards the Age of Aquarius 226

What has changed in the last five hundred years? 228

Astronomical and Scientific Developments 228

Deciphering the cuneiform, and developments about the ancient history 231

Discoveries revealed through dreams, sightings, and visions 233

- Developments on Spiritualism 241
- Other important developments 242
- Sources: 248

SECTION 10

AS ENKI'S AGE OF AQUARIUS BEGINS 251

- All that happened; was it destiny or kismet? 252
- Sources: 264

APPENDIX-1

THE ANUNNAKI IN THE ENLIL CLAN (EAST) 265

- THE MOON GOD SIN/ NANNAR/ EL 266
- OTHER TEXTS ABOUT INANNA 273
- SHAMASH: THE GOD OF SUN 275
- NINURTA: THE ASSYRIANS' ARCHER GOD 278
- The Myth of Anzu and Ninurta 279
- Was Lilith the very first feminist of history? 281
- Sources: 285

APPENDIX-2

THE ANUNNAKI OF THE ENKI CLAN (WEST) 287

- THE LORD OF THE AGE OF THE RAM: MARDUK/AMON RA ...288
- THE GOD OF NUMBERS: THOTH 289
- THE PRINCESS OF THE ENKI CLAN: ISIS 291
- THE FATHERLESS CHILD: HORUS 294
- THE SOUTH AFRICAN GOD: NERGAL 295
- THE MINING GOD: GIBIL/GEB 296
- Sources: 297
- Bibliography 306

PREFACE

I have lost my father, who pushed me to be a researcher, while this book was in the process of being written at a considerably early age. This unmatched man who taught me to question life and death, and who made me the Gok Turk that I am today, lead me to experience two feelings I had perhaps never tasted before; pain and missing someone, by his unexpected death at the age of sixty. Just like millions of others who have gone through similar pains, the same questions kept on lingering in my mind over and over again during the early times of our separation:

Why did death exist? What was happening to consciousness while the body was going into earth?

Did the soul really exist? If it did, where did it go after it was separated from the body?

Was it possible to beat death, to overcome fate?

While all these questions were lingering in my mind, the subject of "Luqman the Wise", which my father had told me when I was 10 years old, came to my mind once again. This narration was perhaps a story, or maybe a symbolic telling of the realities, but it was possible to observe human beings' endeavors for not dying in all periods of the history since the Epic of Gilgamesh.

In the Epic of Gilgamesh, Gilgamesh whose father was a human being, and mother a goddess (Anunnaki) would consider himself as a half deity (half Anunnaki). However, he also knew that this would not save him from dying. The question Gilgamesh asked to Utu, the Sun God (Shamash) before embarking on his journey for the search of immortality, perhaps was the most important question ever for the human kind:

In my city man dies; oppressed is my heart.

Man perishes; heavy is my heart.

The tallest man cannot reach the heaven.

The broadest man cannot cover the earth.

I peered over the wall,

I saw the dead bodies...

As for me; I, too, will be served thus?

Is this how my destiny will be like?

While Gilgamesh's cries have been prevailing the last five thousand years, since that day several other historic figures have been chasing immortality. Was the searches of the Persian King Cambyses, Alexander the Great of Macedon and thousands of others for the Fountain of Life in order to postpone death really meaningless? Or, would the human kind reach this goal when science and technology advanced, and the science of genetics was decoded?

Immortality, which has been chased by human beings for the entire history, has gained a new dimension nowadays due to the advancements in technology and science. The modern Gilgameshes and Alexanders of our era are dedicating some of their wealth at hand to the research being carried out with this goal. This quest, which manifests itself in many ways from brain transplant surgeries to setting up a colony on Mars, and from consciousness transfer experiments to other experiments pushing the boundaries of science, perhaps, soon enough will end with a victory against death. Let us go even further than that; although our scientists are yet to succeed in this, maybe there are people out there who have already succeeded in doing it.

The long lifespans of the gods and goddesses of polytheistic religions have made us deem them as immortal back in the days. We actually think that these gods and goddesses whom are referred as "The Anunnaki" in Sumer are extraterrestrial, and we observe that the works they have carried out in the name of extending the lifespan after their arrival on earth have just begun in our scientific world.

Since the day I have started my research I have tried to explain through the ancient tablets that have been acknowledged by the universities around the world, sources that are formed of respected researchers, scientific research and numerous articles, ten folds of education and seminars that what is missing in the history of the world is the "Anunnaki variable". I was still putting an effort for the same when this book was ready for publication in May of 2018. In the meantime, my principles remained all the same. In the time elapsed so far, I have not shared any information that transmits negative energy. In every single work of mine, I tried to reflect my own point of view with a sincere and positive attitude. Our field of subject does not have a certainty and clarity. That is why I have always left the empty half of the glass to the reader. I tried to present different points of view in line with the principle of information based on proof and sources while improving myself from one day to another, and never deviating from my course an inch.

Lately, we observe that the Anunnaki are the most talked about extraterrestrial beings. As that is the case, some misunderstandings, wrong information and channel information turned to the Anunnaki. Thus, one of the goals of the book is to share the information about the Anunnaki that is within my reach as is with the readers, and to be a source for the most accurate information on this subject. You will find everything about the Anunnaki in this book. Furthermore, you will come to see that our modern world is actually a work of them.

The second subject the book covers consists of the relation between the information revolution that has been going on for the last five hundred years and the Anunnaki. The developments in the fields of writing and history alone constitute an unparalleled phenomenon. The countries that mobilized the archaeologists since 1800s, and which deciphered the collected data and the tablets correctly, have each turned into a superpower nowadays.

What was revealed in the tablets that the leading governments precipitated the sharing of the ancient cities of the Ottoman Empire? Why did the British excavate in Mosul, and the French in Khorsabad while the Russians excavated in Babylon, which was among the cities that diverted the flow of history? Why did these nations never interfere with each other's excavation site, and have always kept the data they have gathered to themselves alone?

Even 150 years after the pillaging of these ancient cities, why was storming the Bagdad Museum, and carrying away fifteen thousand pieces was among the first things USA did in 2003 during the invasion of Iraq? Why were the cuneiform tablets, which were at least six thousand years old, and which were telling mostly about the Sumerian history, were stolen in their entirety? Those who have carried out this pillage were not the ordinary people who were just passing by, and who simply took stones from the ground and smashed the display windows. These were people who wore headphones, who worked in coordination, who were able to pass through double-locked gates, and who knew the secret sections of the museum.

The Sumerologist Jean Bottero mentions in his work "Mesopotamia" that the tablets, documents recorded on stone or metal, clay tablets and all the other documents are distributed to the great museums of the world. According to him, there are half a million pieces collectively at museums such as the Iraqi Museum in Bagdad, the British Museum, Berlin Museum, Louvre Museum, Istanbul Archaeology Museum, Brussels Cinquantenaire Museum, L'Ermitage de Saint-Petersbourg Museum. So, why are all these pieces important?

The answer to this lies in the fact that the tablets have an enormous amount of information. Among these are the information on DNA, the positions of gold mines, the notion that the deities of the polytheistic religions are actually extraterrestrial, the mysteries of the entire mathematical system, initiation knowledge, the cosmos cosmogony "la mudu", and numerous others. Giant gold corporations such as General Corporation had read in the information on the tablets that there were four hundred thousand years old gold mines in Abzu. Where was Abzu? It had to be discovered. Eventually, this region was identified as South Africa, and the mines were discovered. Besides the gold, the most significant diamond reserves of the world were discovered, too. The gold of Assur, Babylon and Akkad which rested in the Mesopotamian temples made Europe rich for a second time following the gold of Mayans, Aztecs and the Incas.

If we take into consideration that the deities the American indigenous people worshipped, and the rituals of gold presentation to them were almost identical with those of Assur, Babylon, Akkad and etc. this question comes to our mind: "Why aren't there any gold in the Mesopotamian temples?"

In his work called "Babylon", based on the information excerpted from Herodotus, Joan Oates has written that the Chaldeans presented the temple with two and a half tons of gold daily during the Bel-Marduk Festival. Where is all this gold now? Neither the Russians excavating at Babylon nor the other nations excavating at the other sites the last two hundred years have ever mentioned about the temple gold found. Maybe they have carried away what they found in the ancient cities' temples in secrecy. Under these circumstances, one can understand the reasons for the Germans, the British, the French and the Russians sharing these cities among themselves. However, consequently, a much more important question lands: Is the hidden heritage of the Mesopotamian nations which is now pillaged, and which remains out of Iraq and Syria, in Turkey? If there still is a hidden heritage in the unexcavated cities of South Eastern Anatolia, then, is there still any gold in the hidden temples on top of which tomb-monuments such as Balikligol (Pool of Abraham) are built?

While Europe was getting richer with the Mesopotamian excavations, and the humanity was facing its real history in the 19th century, after about one century later, in Turkey as the inheritor of the Ottoman Empire, a sun was born: Ataturk. While the War of Independence was being fought on one hand, he asked the "Works of Antiquities Directorate" to be founded under the jurisdiction of the Board of Education on May 9, 1920. Later on, this office was renamed as the "Directorate of Culture".

In this race in which we had long lagged, archaeological excavations had been initiated in many parts of our country under the supervision of the National Government. In 1933, Sumerbank was founded with the purpose of making the name Sumer to be familiarized and long-lived. In the same year, the very first students were sent abroad to get archaeological education under government scholarship. Ataturk asked Sedat Alp through Reshit Galip to get "Hittitology" education. Sedat Alp who got his education at the Berlin University between 1936 and 1940 became the very first Hittitologist of Turkey. In the same manner, our world-renowned scientists Ekrem Akurgal and Halet Cambel were among the students Ataturk had sent abroad for education on archaeology. Ataturk did not only send students abroad but also made an additional article to be included in the regulations of the History Foundation, which had been founded on April 15, 1931: "To send committees of research, excavation and discovery to where it is necessitated in order to get the documents and materials to shed light on the Turkish history."

In the period following this, he had been influential in fostering Sumerologist such as Muazzez Ilmiye Cig and Mebrure Tosun by inviting world-renowned Sumerologist such as Benno Landsberder based on the recommendations of Sedat Alp.

Unluckily, the progression did not continue during the early years of the Republic. Ataturk passed away. In the post-

Second World War era, the superpowers shared the ruling of the world. What was left to our share was third grade rulers and a third grade civilization.

It is a fact that our civilization, which had been in a continuum of decline in terms of information after the Sumerians, has developed under the leadership of the West in the last five hundred years. Why have science and technology passed into the hands of the West, which has been standing still the last two thousand years, and which has no preparations whatsoever as we know? Human beings have been living without any access to information almost for five thousand years. They could as well have lived for another five thousand years in tents, washing their clothes by the rivers; thinking that the earth is flat, and that the sun is orbiting around the earth. These are the questions this book aims to find answers for.

In this book, we have told about and familiarized the Anunnaki, and looked into the contacts which we think special people like Enoch, Admiral Byrd and Rumi established with them. Through them, we have tried to decipher the messages given to the awakened people of our day and age who are referred to as "The Elect" and "The Righteous".

Each section of the book has been written independent from each other, and the sources that were referred to have been listed at the end of each section. And at the very end, all the other sources have been listed altogether. As a high volume of citations have been taken from the books of the researcher-author Zecheria Sitchin, only those parts cited from his works have not been especially indicated.

In several parts of this book the term "mainstream science" has been repeated because when one speaks of mysteries that are still not resolved by contemporary sciences, and the problems they refrain from accepting, one is obliged to differentiate between the two after a point. Based on Malinowski's definition, science is "a body of rules and conceptions based on experience, and derived from it by logical inference, embodied in material achievements and in a fixed form of tradition, and carried on by some sort of social organization". In other words, science is conservative, and from time to time this conservativeness harms itself. It rejects new concepts with severe reactions, and sometimes it remains wrong. Nikola Tesla's ideas, which were ridiculed at his time stand as the greatest example for this. It is not the purpose of this book to criticize science but rather to remind its conservativeness and lacking parts on certain subjects.

In the upcoming sections, the reader will come across with various different comments and points of view. Some of these points, they will find ridiculous or impossible. With others, they will even go further, and deem them as wacky, and disregard them. On some, they will linger inevitably, and an "I wonder..." will rise, and he or she will comment on the comments. Eventually, the reader will decide on the actuality of what is being told. The real purpose of this book is to push the reader into a new way of thinking, and disturb him or her where necessary. For those readers who are extremely confident of their knowledge, this work of ours will not be too meaningful.

When Planet X/ Nibiru is announced as the outermost planet of the Solar System soon, when history is re-written, and the Anunnaki are accepted as a "fact" not as part of "mythology", we will have a clear conscience for having fulfilled our duty by presenting the information to the public previously.

SECTION 1

WHICH ONE IS THE SIMULATION: THE HUMAN BEING OR THE UNIVERSE?

We must always be persons seeking the truth, and only when we find it and become convinced that we have found it, dare to express it.

Mustafa Kemal ATATURK

The Limitations of Junk DNA and Simulation

There is a movie called "The Thirteenth Floor" which was directed by Josef Rusnak, and released in 1999. The interesting content of this film has to do with the question we have imposed. According to this movie, even all the probabilities of our daily plan in the routine of our lives have been calculated. The decisions we make with our free wills within these probabilities can only extend as far as deciding on which streets to take to cover a certain route. The small details that differ from one individual to another are subject to a system of their own, too. In short, everything we do within a day, in fact, are pre-determined, pre-calculated, and run within a system. Even the things we decide to act upon on the spur of the moment are all within the scope of the probabilities. The simulation within which the individual exists is founded on this system. Regretfully, although an exit from the simulation is improbable, it is quite possible to observe its existence. The only thing we need to do for this is to go somewhere which is not in the calculation of probabilities for that specific day when we wake up one morning.

If I were to give an example from my own life; when I wake up to a routine work day in Istanbul, there is no probability for me jumping in my car, and going to Polonezkoy. The movie "The Thirteenth Floor" directs the audience to go to an improbable place just like me waking up on a work-day morning, and going to Polonezkoy. According to the movie, when you begin such a journey, the system seeks to make you give up on this journey, and presents many obstacles. It emphasizes that you will have gone beyond all probabilities that have been planned for you, and that you will have witnessed the real life as is when you follow that route without any intimidation, and by overcoming every single obstacle. And then, it hits one like a ton of bricks by presenting the limitations of the simulation for you.

I have not lived through such an experience but several people have been pondering upon this subject, and generating ideas about it. You might have read the news about the two billionaires investing a huge amount of money recently with the scientists in order to come out of this simulation (1). Based on this piece of news which states that many people in the Silicon Valley have been seriously obsessed with this simulation hypothesis, and that they have disclosed that they have adopted the belief that our perception of reality has been produced in a computer, the two billionaires have gone to the extent of secretly coming together with the scientists in order to free us from simulation. Elon Musk, allegedly one of the two billionaires in this piece of news, says "All of us are living in a game that is being controlled by others almost entirely like in the movie Matrix.", and he adds "There's a billion to one chance we're living in base reality." (2)

To get a definite answer for the question "Which one is the simulation: the human being or the universe?" we must get to know the two variables, that is; the human being and the universe very well. However, we do not have enough information on either of them. All we know about the universe consists of nothing but theories and the images taken by the satellites and telescopes we have sent to the orbits. What we know about our Solar System is somewhat more but even that is not sufficient. The information we have so far about the Earth on which we live is not at a satisfactory level either. We are in the 21st century, yet, where there are people who are still discussing whether the world is flat or round, we cannot say "The Universe is a simulation, and we are living in this simulation." based on what we know of the Earth, the Sun System and the Universe.

Although we have advanced somewhat more on what we know about the human beings, our second variable, through the effort of modern medical science, what we do not know about the human body is still magnitudes more than what we know about it. The research carried out on "Junk DNA" or "Garbage DNA (Noncoding DNA)" lately may take us one step closer to the solution. (3) To explain it briefly; only %3 of the genes in the human genome are coded genes, and can create protein. % 97 of the genes are none coded, and cannot be converted to protein.

Thus, these are named as "meaningless genes". The strange part of it all is that the meaningless genes are controlling the meaningful ones. The continuous synthesis of telomerase which is known as the long-life enzyme is also hidden in these meaningless genes.

Some scientists such as Prof. Sam Chang claim that these meaningless genes also known as "Junk DNA" were transformed into us by some type of an extraterrestrial programmer. (4) On the other hand, Russian biophysicist and molecular biologist Dr. Pjotr Garajajev and his colleague Vladimir Poponin, have stated that the blockages in these genes can be eliminated by X-rays and radio and light frequencies beamed from a distance. Furthermore, they have discovered the vibrational behavior of DNA. (5) They have proposed that DNA can be affected with words and frequencies, and be re-programmed without cutting and taking out the DNA and without altering the genes one by one.

This allegation put forward by scientists makes one infer that this blockage in the genes of human beings can be eliminated by individual efforts, and also that our genes can be manipulated by others. Of course, this is not an easy task because based on the research that was carried out, it was determined that only very few of the genes which were freed from blockage were related to cancer. The relation of these freed genes to cancer brings to our minds the existence of a precaution taken against the possibility of eliminating the blockage from the entirety of the genes.

This method can be labeled as a mine field. According to researchers and scientists who believe in the existence of "Ancient Astronauts" just as we do, there is a design that obstructs us from activating our genes in their entirety. We hope that when the time arrives human beings will lift this blockage completely with all the dangers being eliminated and reach his or her real power.

This is the actual question: what are the secrets hidden in these junk genes at a rate of %97? Do these junk genes stand as one of the most important factors which separate us from facts and make us think that we live in a simulation? Maybe we will reach the real human being when we activate these genes completely. We will see areas we have not seen before and hear frequencies we have not heard before. Most importantly, we will be able to live much longer than we can now or our sixth sense or third eye will open. It seems like we will not be able to get answers to these questions in the near future either, but there is one thing we know for sure: without finding out what is inside these genes we cannot say that we have figured out the being which we call a human.

Dr. Chang's thesis on extraterrestrial intervention on human beings calls for this question: If it did happen, when could it have happened? On this token, we have to indicate that this aforementioned extraterrestrial intervention does not point out to creating or being created. In other words, it is not about a divine touch. It is the technological intervention of a different hand than ourselves that is far more developed on Homo Erectus which already existed. We have been similarly applying in vitro fertilization method for many years already on humans.

In order to have an idea on when the Homo Erectus could have been intervened, let us have a quick look at the history of the modern human being (who is a stranger to the world by all means), in other words, Homo Sapiens Sapiens.

History of Homo Sapiens Sapiens

The Sun System comes into existence 4,6 billion years ago and 4 billion years ago the biological evolution starts on earth. According to the Sumerians the latter date stands as the date when one of the satellites of Nibiru collides with the main planet Tiamat and divides it into two and one half forms the Earth.

During this period, the seed of life was carried on to the Earth and primitive life without oxygen begins. (6)

2,5 billion years ago primitive life transforms into multicellular life. The first bacteria comes into being.

500 million years ago, the earliest animals, shelled creatures come into being. The creatures of the seas begin going on land.

250 million years ago dinosaurs, giant reptiles and giant plants come into being.

65 million years ago life on Earth vanishes following a meteor strike.

25 million years ago the earliest human-like creature comes into being.

14 million years ago the hominids come into being.

3 million years ago the first being classified as homo comes into being.

2 million years ago the first being, resembling a human, lives in Africa. (advanced Australopithecus, Lucy)

1 million year ago, Homo Erectus comes into being.

200 thousand years Neanderthal human comes into being. (The cave man)

Although more than 2 million years had passed between the advanced Australopithecus and Neanderthal, the sharp-edged flakes of stones which were used as tools by both groups are almost identical and the appearances of these groups can hardly be distinguished. (7) In other words, although such a long period of time had passed, there were no developments. 28 thousand years ago, the Neanderthal human breed becomes extinct. If the Neanderthal breed became extinct, the prioritized subject of wonder is: "Who are the ancestors of the Homo Sapiens Sapiens?"

As the research goes on, it gets discovered that modern humans named Cro-Magnon who resembled to us quite a lot, came into existence almost out of the blue some 35 thousand years ago and that they co-existed and lived together with the Neanderthal for 7 thousand years. The mystery around the Cro-Magnon human's coming into existence increases even more with further discoveries because Cro-Magnon has evolved from an early Homo Sapiens who lived in Africa about 250 thousand years ago. (8)

Actually, it is irrational that the modern humans came into existence 700 thousand years after the Homo Erectus and 200 thousand years before the Neandertal humans. According to Prof. Dr. Theodosius Dobzhansky who is an authority on this matter, the modern humans have many relatives who are from the same line of race in terms of fossils, but they do not have ancestors. Civilization has started in Africa, but the question that is yet to be answered is this: "Why did civilization come into existence? Based on all the data, in fact, the humans should have been still lacking civilization. There is no obvious reason for us to be more civilized than the primitive tribes. However, the true puzzle lies not with the undeveloped Bushmans, but rather in our progress. (9)

2 million years had passed between the times when the humans started cutting stones naturally and the times when they started cutting and shaping these stones pursuant to the purposes. Why did not another 2 million years pass to learn to use the other tools? How come astronauts landed on the Moon 30 thousand years after the Neanderthal humans?

It is exactly at this point where science comes to a deadlock that it is being suggested through the Ancient Astronauts Theory that the Homo Erectus had an extraterrestrial intervention around 250 – 300 thousand BCE and that the Homo

Sapiens was created by tampering the genes of the primates, utilizing a method similar to in vitro fertilization. Although we do not know who made this come true, according to the Sumerians, these were the Anunnaki.

We think that the junk DNA, too, have come to existence through the technological intervention of the Anunnaki. Even though there were other interventions back in time, the ancestors of the contemporary humans came into being through this very first intervention. Nowadays, there is another intervention that is being applied on our DNAs by using different tools like GDO or Chemstrail. Consequently, the next generation will be a different type of human kind.

We see that the answer to our question starts shaping up gradually and make such an assessment: If we have information only about %3 of our genes and if we can only come up with ideas for the remaining %97, could we be possibly living the reality? If some others have left these genes unfunctional, we wonder what we will be facing when we make these genes completely functional in their entirety once again. Are we going to go back to Homo Erectus? Or else, are we going to awaken to the reality becoming superhumans?

Researcher, author Gregg Braden tells that we only use 22 out of 64 genes in our DNAs basing it on his cross-verified genetic research and questions what would happen when we can access all 64 of the codes. He adds that our DNA has been most probably genetically manipulated by the Anunnaki and then he asks "Why on earth some entities out there limited our genetic capabilities by switching off most of our DNAs?" (10)

Based on the research by another researcher, author, Robert Morning Sky in fact our DNA is superior to the Anunnaki race because we also have the AKU DNA which gives us our consciousness. According to the author, AKU is the part that has been transformed to the human by an extraterrestrial race other than the Anunnaki. This is known as the "Feather of the Gift" and it is the very reason why we feel something magical when we come across with a feather randomly in our daily lives. (11)

Although the science of our day and age cannot give the answers to all these questions, we may very soon reach some answers because in the last five undred years the curtain in front of humanity has been gradually lifted and we take new steps daily on the route to reaching information. Individual and societal developments both speed up constantly. Right at this point, there comes another question to one's mind: "Why did information begin flowing to the humans only in the last five hundred years? How did we reach this information? Which methods were used?" Before we dwell in the field of answers to these questions, we need to have a look at how our science followed the traces of the Anunnaki.

Micro and Nano Robots

At present, just like many other tech companies <u>SRI International</u>, too, develops micro-robots. Each mini robot has its own tools and a few robots can come together and produce macro size products. These mini robots are actually being controlled by small magnets and a magnetic field formed on a circuit board. That means these mini robots are produced by using low-cost materials and they are suitable for mass production. SRI has also come up with a way for these mini robots to self-produce these tools they need. As all these micro robots are fundamentally same they can utilize their tools to achieve various different jobs. Furthermore, they have the capacity to produce tools for other robots by forming their own tool ateliers.

According to Ron Pelrin, SRI's chief scientist for the robot program, these robots can be integrated with larger robots and can act like a robot immunity system. They are at a level where they can observe, maintain and if necessary, fix the larger robot. (12)

There are robots used in the field of health, too. These resourceful robots equipped with the great advantages of being small can be taken in a body through the mouth in the form a soft capsule. These small robots which can be used in carrying medicine can carry out an endoscopy when necessary. Besides, through the use of special blades placed within them they can cut out a desired section in the body. Thus, they breathe new life into the biopsy applications. These robots which can make surgical interventions demonstrate already that this adventure which was initiated with an inspiration from the nature will take us to limits beyond our imagination. (13)

As the micro robots are being used prevalently in various fields, they hold an important place in our lives. However, after this point in time, what will intrigue our minds is actually the nano robots otherwise called as nanites which were introduced into our lives in 2016. These robots which cannot be distinguished with the bare eye and which are one thousand times smaller than the width of one single human hair will be the architects of the future. (14) Berbard Feringa, Jean-Pierre Sauvage and Sir. J. Fraser Stoddart were awarded the Nobel Prize in Chemistry in 2016 for building these "machines" in nanometric dimension. (15)

With the contemporary technology we can place one hundred thousand silicon circuits to the head of a pin. Micro and nano technology are being utilized almost in every field nowadays. While micro-electro-mechanic systems technology is being used in electronic equipment, bolometers, gyroscopes and detectors of humidity, temperature, pressure, tremors and shakes, we come across with nano technology mostly in the units of electronic circuits, batteries, smart textile and paint materials and automotive, aviation and space industries. (16)

Simultaneously, micro and nano technology paves the way for discoveries pushing the limits of the human imagination for the future. In the light of all this information we can easily concur that chips, nano technology and this technology's products nanites will re-shape the future. As science always advances humanity, one can ask "What is so bad about it?" Science means advancement, yes, but when we consider examples such as the bedeviled arms industry, the viruses and the nuclear technology right in front of our eyes, we are justified in worrying a little.

Getting inspiration from the article my dear friend physicist Meral Coskun penned for the periodical Gaia, let us comment on these developments: (17)

At present chips are being used both in animals and humans. While the chips have been in use for many years to stop the animals from getting lost, nowadays chips are also being used as a precaution against kidnapping of the affluent people in countries such as Argentina and Brazil. In 2015, a technology corporation in U.S.A, Applied Digital Solutions have applied to the government in Florida for permissions to inject chips into humans. And in 2016, at Tictail, a corporation in Sweden, the process of placing chips under the skin has commenced so that the employees can use the photocopy machines, pass through password protected doors and to shop using their account defined to their chips.

Other than these chips which are as big as a single rice grain, there are much smaller chips which can be placed in the neck and inside the optic nerves. These are already being used on soldiers as a remote control.

In 1946, in USA, chips were placed in many newborn babies without the consent of their families. In 1950s these chips went through trials on animals and humans. Several researches were conducted on controlling their behaviors as well as their body and brain functions. Until 1970s the chips could be detected through X-ray films. After the commencement of the use of gallium arsenide instead of silicon chips could be made much smaller. In 1973, the assassinated ex-prime minister of Sweden, Olof Palme had decided that chips must be placed on inmates and this permission was included in the government reports of that period.

A chip of 5 micro millimeters or even smaller nano robots (nanites) can even be taken into the body by breathing in, injected with the medicine and transferred into the body by putting them in the water, the food, the toothpaste and the shampoo we consume. Such chips loaded with a program and which fulfill every order given were placed on the space probe New Horizons which was launched to conduct a study on Pluto in order to direct and control it. In this way, the probe which is billions of kilometers away can easily apply the orders given from earth. We observe that we can make a space probe functional by programming it like this. So, why should it be difficult to control the humans by placing nano scaled circuits into the human body?

Alright, so, what could the humans be made to do with these nano robots or chips?

The technology products nanites which are even smaller than the tip of a needle can alter the complete structure of the human being. After a while orders given to these nanites and our bodies can engage in a war. By the help of these chips monitoring and controlling humans from a distance can be managed. Furthermore, human beings can be made to take various actions through electromagnetic wave transmissions.

Much more importantly, new software can be uploaded remotely into the codes in our DNAs. Those who have this technology can create a new race or a new culture as they wish. They may be doing it in such a way that we may not be aware of it.

When we look into the contemporary perception of shopping among human beings we can easily observe the change. How did "I need this pair of shoes." evolve into "I deserve better shoes." or "I will spoil myself a little." lately? How did we turn into a world where people are on the verge of committing suicide because of their debts for the credits and the credit cards the banks give out as if giving candy to the kids? Every eight person out of ten around us are struggling with debt. Well, have we all come to this because we are carefree or could there be other factors such as nanites which move and shape our inner instincts?

All these are not conspiracy theories, but rather the realities of the present and even the future. If we say that multifold higher of the technology we have reached at present already existed with the Anunnaki who came to Earth 445 thousand years ago, do you think it would have been very difficult for them to tamper with the genes of Homo Erectus and turn it into Homo Sapiens utilizing a much superior technique than in vitro fertilization?

One Hundred Spectacular Days God Granted

There is a TV series called Stargate SG-1. You must have watched or heard about it. Nevertheless, let us give a brief explanation about the series for those who have not heard:

The first film of the series, Stargate starts with the discovery of a mysterious ring during the ancient Egyptian research in 1930s. There are numerous symbols around this ring. However, no one can decipher these symbols. Therefore, what the ring serves for, the language of the symbols and what kind of element the ring was made of cannot not be solved for a long time. At this point Dr. Daniel Jackson, an ancient languages expert who had been out casted by the scientific world steps in with his interesting theories. Daniel Jackson's theory constitutes the absolute foundation of the movie. According to this theory, the deities of the ancient mythology are extraterrestrial, and the pyramids are the landing runways for the spacecrafts of these extraterrestrial beings. We can frequently see the ancient deities such as RA, THOR, APOPHIS, ANUBIS, HATHOR, SETH in the series anyway. The adventure which started as a movie continued with three separate movies and three different television series lasting fifteen seasons. (18)

Now I want to talk about the "Brief Candle" episode of the Stargate SG-1 series which was broadcasted in 1997 which falls into our main subject matter. In this episode, our team goes to an unknown planet and in a surrounding, which looks like the ancient Greek temples comes across with beautiful people who attract our attention with their ancient Greek style clothing. Here the women are perfectly beautiful, and the men resemble the ancient Greek statues. At the temple of the city there is a statue of a deity and the public presents its gratitude daily to this deity. Why? Because this deity whose statue they have erected has granted them one hundred days each of which is superior to the previous one.

Initially our team do not get it. Yet, when they see that a newborn baby grows up in just two days and becomes a young girl in five days, they begin grasping the issue. They are literally facing an accelerated life and people who can live for one hundred days. Then, when one of the members of this society who goes to sleep to rest each night only to wake up to fun and entertainment in the morning has sexual intercourse with someone from our team things happen. This person starts getting old all of a sudden. After lengthy research the answer is found: the nanites.

Once we look at the result we see that the "extraterrestrial being" whom they see as their deity has carried out an experiment and re-written the DNAs through these nanites placed into the human body. However, these nanites must have a triggering factor, a stimulator. In other words, the nanites cannot function without a main computer. And, sure enough, that triggering computer was smartly hidden in the statue.

Reflecting upon our own lives, let us think of the one hundred days of the movie as hundred and twenty years, the maximum humans can live. Perhaps we used to have a lifespan of fifty thousand years and this timeframe was reduced to one hundred and twenty years through such an intervention. Consequently, we got cut off from space and imprisoned in the Sun System with these short lives. Can it not be? When we glance into the past, do the Sumerian King Lists, Egyptian King Lists and the Holy Books tell us that there are humans who live for thousands of years or not?

The Great Code in the Junk Genes

Let us see how Prof. Sam Chang, who studies junk genes, considers this matter.

"If we think of about it in our human terms, the apparent "extraterrestrial programmers" (according to us, the Anunnaki) were most probably working on "one big code" consisting of several projects, and the projects should have produced various life forms on various planets. They have also been trying various solutions. They wrote "the big code", executed it, did not like some of the functions, changed them or added new ones, executed again, made more improvements, tried again and again.

The apparent "extraterrestrial programmers" may have been ordered to cut all their idealistic plans for the future when they concentrated on the "Earth Project" to meet the pressing deadline. Very likely in an apparent rush, the "extraterrestrial programmers" may have cut down drastically on the big code and delivered basic program intended for Earth."

However, perhaps they were not sure at that time about which functions of the big code would have been necessary in the future and thus they left them all. Instead of cleaning the basic program by erasing all the errors of the big code, they converted these to comments and in a rush forgot a few symbols of /* here and there in the comments. Consequently, they presented humans with the illogical expansion of cells which we know as cancer.

There are three options to the problem. Either delete all the /* symbols and comments and clean this way the basic code, or add all the missing */ and avoid illogical mixing of the basic code with the big code. Alternatively, in the third option, remove all the / symbols and let work the basic code with the big code as a complete program. Unfortunately, none of these options are within our capacity. If we were able to efficiently insert genes into the chromosomes of living man, our breakthrough discovery would mean instant cure for all future cancer cases; at least from the programmer point of view. Theoretically, we can do it in a laboratory, but we have no practical means to implant the repaired DNA into living subjects. The mystery of "junk DNA" and cancer seems to be solved, but no quick cure shall be expected. The best thing we can do now is to try nourishing new, cancer-free line of humans with gradually debugged basic genetic code. That will take a long time. Four our children and us, there is no hope on horizon.

Sooner or later, we have to come to grips with the unbelievable notion that every life on Earth carries genetic code for his extraterrestrial cousin and that evolution is not what we think it is."(19)

Maybe the genes Prof. Chang believes to have been left after an intervention in a rush were formed deliberately and voluntarily. Nowadays, the notion that the change over from Homo Erectus to Homo Sapiens was made in the laboratories and after various experiments is accepted by many researchers. Those who would like to learn this process in detail can read The Lost Book of Enki by Zecharia Sitchin which is an anthology put together in the light of the ancient tablets.

If all these propositions are correct, this is the result: A programmer who is much more advanced than us on the subject of nano technology has made it possible for us to reach today by tampering with our genes several times. The first intervention was made somewhere in the period between 250 thousand and

300 thousand BCE. According to the Sumerians the Anunnaki are to be held accountable for this intervention.

Right! So, are there any other races other than the Anunnaki who have visited our Earth? In fact, at present we see thousands of broadcasts and publications which mention the extraterrestrial races. And there are many writings which mention races coming from various stars and contacting the earthlings.

There must be trillion types of extraterrestrials in a universe where there are billions of galaxies and trillions of stars. However, when we look into history, we come across with only one type which we hold accountable for intervening with the humans and whom have left thousands of traces and signs as proof that they have visited the Earth. This type is nothing but the Anunnaki whom are known to be the deities of the polytheistic religions. (Figure 1)

Figure 1: Ancient Astronauts: The Anunnaki

Sources:

1. http://www.bizsiziz.com/iki-milyarder-insanligi-dev-bir-bilgisayar-simulasyonundan-cikarmak-istiyor/
2. http://www.sabah.com.tr/dunya/2016/06/05/yasadiklarimiz-bir-simulasyon
3. https://en.wikipedia.org/wiki/Noncoding_DNA
4. http://www.kosulsuz-sevgi.com/dna/insan-dnasinda-dunyadisi-genler/
5. http://www.rexresearch.com/gajarev/gajarev.htm
6. https://www.bibliotecapleyades.net/hercolobus/esp_hercolobus_34.htm
7. SİTCHİN, Zecharia, 12. Gezegen, Ruh ve Madde Yayınları, İstanbul, 2001, S. 13
8. SİTCHİN, Zecharia, 12. Gezegen, Ruh ve Madde Yayınları, İstanbul, 2001, S. 14
9. http://blog.milliyet.com.tr/ilkel-busmanlar/Blog/?BlogNo=120969
10. http://www.abzu2.com/tag/anunnaki/
11. http://in5d.com/anunnaki-message-published-in-ufo-magazine-in-1958/
12. http://superileri.com/birer-minik-robot/
13. http://www.bilimgenc.tubitak.gov.tr/makale/buyuk-isler-basaran-kucuk-robotlar
14. http://www.muhendisbeyinler.net/nano-malzeme-nedir/
15. http://bilimfili.com/2016-nobel-kimya-odulu-nano-olcekli-makinelerin-oldu/
16. http://www.nobelprize.org/nobel_prizes/chemistry/laureates/2016/press.html
17. http://sanayilesme.ssm.gov.tr/ARGE/MUKNET/MikroVeNanoTeknolojiler/Sayfalar/default.aspx

18. http://www.bizsiziz.com/nano-boyutta-kolelestirilen-insanin-psikotronik-cag-teknolojisi-ile-olusturulan-yabak-hali/
19. http://dizibilimi.blogspot.com.tr/p/stargate-film-serisi.html
20. http://www.kosulsuz-sevgi.com/dna/insan-dnasinda-dunyadisi-genler/

SECTION 2

WHO ARE THESE ANUNNAKI?

You have to know the past to understand the present.

Carl Sagan

Ancient Ancestors; The Anunnaki

Anunnaki is a Sumerian term. Although in general it gets to be used in the context of "those descending from the Skies to Earth", in fact, it means "An's successors". Although at present the Anunnaki are known as the Sumerian deities, the Sumerians had actually never worshipped them in the sense we know. They had only served them calling them as "Holy Masters". In return, they have asked from them protection, knowledge, healing and power. In the past of almost every culture, one can see gods and goddesses resembling human beings with various jobs, but upon closer look into details, it is understood that they all originate from Sumer. No one has ever come across with any records, lineage registries or tales anywhere mentioning any deities older than the Sumerian ones. These deities have been named in the original Sumerian forms or in the latter Akkadian, Babylonian or Assyrian forms and they have been listed one after another. The list extends with hundreds of names.

When we bring all these deities together, it seems like a chaos formed of thousands of deities. However, once the stories in different mythologies are carefully examined, one can see that the situation is not actually so complex. Because different deities of different societies are actually the re-named Sumerian deities whose numbers and ranks are certain. True, the stories are meddled into each other because they were spread by word of mouth, but Ishtar, Aphrodite, Venus, Astarte, Ashtoreth, Al-Uzza, Ayisit are nothing but the Sumerian goddess Inanna, only named differently in each different culture.

The gods and goddesses we knew through sources such as Herodotus, Homer, and the Torah and also through the mythological stories of cultures have gained a very different dimension after ancient civilizations were discovered and understood. The written language, expressions, passions, wars, love affairs and decisions of the Greek deities mentioned in the works of Greek literature that survived to the present day show us that they were no so different than ourselves. In the Torah, too, all the societies around the Israelites believed in these deities and furthermore, they would meet in person with their deities from time to time. Once ancient languages such as Akkadian, Sumerian and Egyptian were deciphered, the monuments, the pyramids, the ziggurats, the great platforms, the ruins and the stone tablets of the ancient eras all gained different meanings. It dawned upon the researchers that the real source of the Greek and Torah deities and even the Book of Genesis from the Torah was Sumer. It was observed that the deities of the Sumers, the Akkads, the Assyrians, the Babylonians, and the Hittites had human characteristics just like the Greek ones and the ones mentioned in the Torah. Everything we know have changed once the ancient writings were discovered and deciphered.

Thanks to the Behistun Inscriptions of the Persian King Darius in Persepolis we can now read and understand the languages of the Akkadians, the Assyrians and the Babylonians. Darius had this inscription written in three different languages in order tell how he conquered territories and expanded his country and how he subjected the countries under his domination. Around mid-19th century, the resemblance of the language of Old Persian which was one of these three languages to the Persian language of present was noticed and this ancient language was deciphered. Following this, based on this language alphabets for Elamite and Akkadian, the two other languages in mention, were prepared. Then, once the inscriptions in Akkadian which were discovered in the Mesopotamian excavations were deciphered using the Akkadian alphabet prepared in Behistun we were informed of the content of the tablets.

Also, in another location, the Rosetta Stone which was prepared in three languages; demotic, hieroglyphic and ancient Greek in order to be distributed to three main temples in Egypt was discovered. This stone, which was discovered during Napoleon's campaign to Egypt, is the inscription stone for the Hellenistic Ptolemaic Dynasty ruling in Egypt in the period 305-30 BCE. The researcher, who solved the mystery of the stone, and consequently, the mystery of the hieroglyphs, is Jean -Francois

Champollion who proved that the ancient Egyptian scripts resembled the contemporary Coptic language. The deciphering of the ancient scripts of Egypt gave way to the birth of Egyptology and this in return facilitated solving the mysteries of the past centuries.

In short, we can conclude that the Akkadian and the ancient Egyptian languages were deciphered by the use of the Behistun Inscriptions and the Rosetta Stone. If Greek and Persian languages did not exist in these inscriptions, we would have never learned about their content. Sumerian is the clearest example for this situation because at present there are no languages resembling to the Sumerian language. Therefore, we owe the understanding of the Sumerian language solely to the library of Ashurbanipal, the King of Assyria (668 – 627 BCE). A library had been discovered in the palace of Ashurbanipal in Nineveh in which ten folds of thousands of tablets were categorically named under subject matter, the authors were recorded, and the series were numbered. For those tablets among the ones in this library that were discovered by Sir Austen Henry Layard in 1849, which talked about history, science and the deities a dip note was added saying that these were copies of the previous tablets. Yet, in another section of this library dictionaries prepared in this ancient language for the Akkadian language were found. Had we not discovered these Akkadian-Sumerian dictionaries ordered to be prepared by Ashurbanipal in the Nineveh Royal Archive, we would have never learned the Sumerian language. Unfortunately, we do not have the same kind of luck for Harappan, the Indus language. As there is no other civilization that resembles the language of this civilization at present the language of the Indus Civilizations is yet to be deciphered. If one day we find Harappan-Akkadian dictionaries just like those of Ashurbanipal, or come across with tablets with multi languages including the Indus language on one side, and one of which we can read, then we can decipher it.

The actual big problems were raised for the archaeologists who were spellbound with the grandiosity of the Assyrian and Babylonian civilizations after they deciphered the scripts of these civilizations. Sumerian, Akkadian, Assyrian and Babylonian put together; from the five hundred thousand tablets recorded it has been clearly deduced that a prior great civilization had been established. More importantly, based on the archaeological discoveries and academic studies, which lasted for a century, it has been revealed that the foundations of our present day civilization lies in Sumer. One observes that written language and literature, schools and temples, medical doctors and astronomers, mathematicians, high-rises, channels, ports and ships have appeared here as if out of the blue. Moreover, records from prehistoric periods on extensive farming, advanced mining, textiles and trading, regulations and justice, concepts on morals and theories of cosmology have been revealed.

The Sumerians have indicated the Anunnaki to be the source of all these developments. They have even mentioned that all the cities had been prepared by the Anunnaki and presented to the human beings. (1)

Almost all the information in the tablets have been the subject of scientific acceptance. The parts that have not been accepted scientifically are those where the Sumerians mention about the Din.Gir. beings. What they called as Din.Gir, in other words as "Anunnaki", had been called as "God-Goddess" in the later periods by the Greeks. According to this, these beings came to Earth several hundred thousand years ago from a very far distance and chose Mesopotamia as their new home. They called these lands Ki.En.Gir, "Country of the Guards", and founded the first settlement Eridu which means "Home in Faraway".

The introduction of the Atrahasis tablets, which were written in 1700s BCE in Akkadian, begins with "When the deities were like humans". The reason for this is that they had dug up channels and cleared the banks of the River Tigris on first arrival. (2) These deities who lived very long on Earth eventually complained about the load of work and implemented a series of technological interventions following the human beings' coming into existence.

Zecheria Sitchin who deciphered the Sumerian language with a different perspective claimed that Nibiru, the Anunnaki's homeland is actually a planet. The Sumerians had frequently used and emphasized the statement that those who had founded the first settlements on Earth were in fact astronauts who came from a different planet. And they had explained how they could have reached such a civilization six thousand years before our times in this way: "We built whatever seems beautiful by the Anunnaki's grace."

Are the Anunnaki a Race from Space?

In fact, it was much before Sitchin that we had learned that this race from space, which was known to us as the gods and goddesses for thousands of years, was called as the Anunnaki by the Sumerians. Although we consider the meeting of Admiral Byrd with a being from space in the underground base at the North Pole in 1947 as a contact with the Anunnaki, the USA-Anunnaki contact in 1954 is the one that has been accepted as the first serious contact in the environment that adopted the Ancient Astronauts Theory.

Allegedly, the ex-president of the USA, Dwight D. Eisenhower, have made contact with the Anunnaki on February 20 and 21, 1954. Edwin Nourse, a leading economist, Gerald Light, a metaphysics scientist, James Francis McIntyre, archbishop of the Los Angeles Catholic Church and ex-advisor for Pentagon, and Franklin Winthrop Allen, a renowned journalist were present at this meeting, too. In here, as the space beings resembled the people of the Northern European countries, they were referred to as the "Nordic Aliens". (3) It is intriguing that Admiral Byrd would also mention that the extraterrestrials had spoken with a Swedish accent during his contact.

In the meeting which had convened at the Edwards Air Force Base, 35 kilometers north-east of Lancaster it was not overtly mentioned that the Nordic Aliens were the Anunnaki.

However, there were a few signs pointing out that they were the Anunnaki, and that they were the representatives Enki. First of all, they looked like the Anunnaki who were described in the thousands of cuneiform clay tablets of the Sumerians. We think that the Anunnaki belong to a Caucasian race with blue eyes and blonde or white hair (some Anunnaki are red-haired and green-eyed) which is classified as the Nordics. Secondly, the Nordics appeared at this meeting in such a rapid way; as if coming out of nowhere. We believe that the Anunnaki who have figured out the frequency technology came from one of the earth bases (maybe the North Base).

According to us, these aliens who are named as the Nordics are the emissaries of Enki. They are pro-human, and they are here for peace. There are allegations that the use of atomic bombs was prohibited to USA during this meeting. We also observe in the tablets that Enki did not ever save itself from any help it may provide to humanity, and to elevate us to its corresponding values. Consequently, we think that those who came together with Eisenhower were most probably the team of Anunnaki and the representatives of Enki. (4)

Four years after this meeting, an Anunnaki message was published in the November-December 1958 issues of the Flying Saucer Review Magazine in the USA This original article by Trench who was the editor, writer and publisher of this magazine was retrieved from one of the volumes of the Fantastic Stories (USA) published in November 1947 in the USA (5) The article which begins with "We have always been with you." indicates to us that they had a high volume of information about humanity. For those who would like to read the messages given to the humans by the Anunnaki, writing "Anunnaki-1958" in the search engines would suffice.

The idea that the Anunnaki who are the Sumerian deities are in fact an alien race was now being debated in that period, and these discussions had even pushed Zecheria Sitchin to learn the Sumerian language. It was Zecheria Sitchin, who introduced us these deities whom we knew for thousands of years, and considered as mythology, by using the term Anunnaki through his work called "The 12th Planet", which was published in 1976. "The 12th Planet" and the series "The Earth Chronicles" that followed it have been translated into more than twenty-five languages, and have sold millions around the world.

Queen Nin-Pubai lived in the Sumerian city of Ur during the first dynasty period of Ur (2600 BCE). Her interesting skull made Sitchin think that she had the genes of the Anunnaki. He had requested the British Museum to run a DNA Test on what was left of her, but he passed away untimely without the project being concluded. Following Sitchin's death in 2010, despite the dense pressure from his fans and followers, the British Museum never allowed the DNA testing. Today, Nin-Puabi's skull is almost extirpated, and the physical remains are kept at the London Museum of Natural History. (6)

The critic Michael S. Heiser has qualified Sitchin as "Without a doubt, the most adamant champion of the Ancient Astronauts Hypothesis of the last few decades". And according to the critic and writer Jason Colavito, Zecharia Sitchin is now one of the most famous defenders of the Ancient Astronauts Theory. Only second to him is Erich von Däniken who is the father of the theories which emphasizes only real beliefs. According to certain writers, both Sitchin's ideas and Erich von Däniken's theories had influenced the sect of Realism that is otherwise known as the UFO Religion. Furthermore, writer Mark Pilkington claimed that Japan's Pana Wave religious group had been rooted by Sitchin.

The criticism against Sitchin's works can be categorized in three; the translations and paraphrasing of the ancient texts, astronomical and scientific observations, and the literalism of myths. (7) Whereas, in our country, the only criticism is on his Jewish identity. We think that it is quite normal for pioneering studies to have flaws. However, when the history of the world is researched in the light of the information shared, it will be observed that the best hypothesis put forward so far are built on

Sitchin's commentaries on the ancient tablets. Presenting a rational explanation through the tablets by interconnecting all the mysteries on Earth has been the most crucial hint for us while seeking the truth. One hundred years ago, when Nikola Tesla stood out with his radical views, he first faced the attack, then the ridicule, and finally the ostracism of the mainstream scientists, and had said this in those days: "Let the future tell the truth, and evaluate each one according to his work and accomplishments. The present is theirs; the future, for which I have really worked, is mine."

Nikola Tesla's ideas, which were once underestimated, exist in every field of technology today. That is why we say that just like in every other subject, on Sitchin's views, too, the future will present the most accurate answers.

During a press conference in 2016, Kazem Finjan, Iraq's Minister of Transportation, has stated that five thousand years ago, the Sumerians had built a space port in the city of Dhi Qar and went to space. This supported the claims of Sitchin. (8)

Pursuant to all these findings, we learn that the gods and goddesses depicted with the face of humans are in fact a space species that has come to our world, and that they are called as the Anunnaki (Nephilims-Igigis).

The planet where the Anunnaki came from is not far from us at all. Nibiru, otherwise known as Planet X today, which is the outermost planet of our Solar System, and who is waiting there to be discovered just around the corner, is the home of this species from space. The Anunnaki who were not so different from us as far as physical appearance except for their height and long lives, – because, as far as we can follow, we came from the same seed – were extremely developed in technology. The technology they owned made us deify them.

Lately, a new trend has come into existence supporting that the Anunnaki are in fact reptilians. In the ancient history records there are dozens of species not looking like the humans. However, in our own research, we have not come across with such a race so called as the Reptilian. Consequently, we find the Reptilian discussions based on clay figurines with interesting heads and huge slanted eyes meaningless. Yet, even if they exist, we can say with peace of mind that the Anunnaki are no Reptilians.

On these soft skinned figurines so called as the Reptilians one can observe an interesting creature without any body or head hair, with an elliptical shaped head and weird intriguing eyes. It is an extraordinary characteristic of these figurines that the lower parts show a female vaginal opening crossing over a phallus or transecting it. It is obvious that these figurines, which are either hermaphrodites or genderless, do not resemble to us. Therefore, they do not resemble to the Anunnaki, either. In fact, they look like the weird hominoids depicted on the figurines. If what the people who claim to have met the Reptilians say are true, then, maybe we can say that those smart species of space who have been in contact with us are not humans, but rather their anthropoids or organic robots.

Nowadays, we, too, consider sending robots to space instead of astronauts in order to lower the risks to the minimum. NASA, who wants to form a colony on Mars until 2030, has already begun designing many instrumental equipment for this. Puffer, or the Pop-Up Flat Folding Explorer Robot, which can move around narrow spaces on Earth or on other planets, was developed with the same purpose. (9) Who can guarantee that the next step is not building human-like robots with artificial intelligence, or even further, organic robots?

We can easily observe the effect of the support of media in the Reptilian trend's popularity. There is not much information on Sitchin in the media and the documentaries, and, besides, the movie industry, which produces movies about everything, remains silent about the Anunnaki. On the other hand, Hollywood, who finds merit in showing the extraterrestrials as

"Creatures" all the time, has been giving support to this view indirectly. Right! So, what could have been targeted by putting forward these or similar subjects, which are fueled by fear?

In the mid-20th century, a decision was reached, and a break from the human spiritual development was taken. The Jewish Liberal Science has been accepted as the only science. Besides, metaphysic studies and real history research have been assigned only to individuals and sects, and their capacity to reach the broader public have been restricted. Additionally, for those people who were awakened upon seeing the truth, a condensed information pollution has been created. According to us, the reason for this is to push an individual to reach the crumbs of real information within the information pollution using his or her wisdom, intelligence and reasoning. Then on, the purpose is to initiate the process of awareness, and make one approach the world from a very different angle. In short, this is a test of intelligence, reasoning and courage. This is only one test out of several that are put forward in order to create the humans of the new millennia.

We are of the same opinion with the mainstream science in reference to the Anunnaki being the "Sumerian Deities". The point where we are in conflict with the scientists is that we consider them as extraterrestrial species whereas they see them as imaginative deities. In other words, a dissension is experienced on this gods and goddesses group whose existence is accepted without doubt. However, all other species of space whose existence is claimed to be true depend solely on ideas and experiences.

Sure enough, we should not eliminate the idea that various beings can visit our Earth within this century. However, since the very beginning what we are defending is that the beings which are defined as gods and goddesses in the ancient information are from space, and that they come from Nibiru, which exists in the outermost part the Solar System, and which awaits to be discovered nowadays. And as sources, we take into consideration the ancient tablets, inscriptions, stelai, holy books like the Qur'an and the Torah, history recorders such as Herodotus and Beresos, and various ancient texts.

Today, different extraterrestrial species that are said to have visited the Earth are the subject of other research. There is mention of beings which are neither human nor Anunnaki in the ancient tablets. We can see in several tablets that the Sumerian God Enki carried out similar studies. In the myth "Inanna's Descent to the Underworld", which is one of the clearest narrations, Enki forms two beings to save Inanna. Ereshkigal must have seen such beings for the first time; based on what is written on the tablets, she asks them "Are you from Earth, or Anunnaki?" These beings have names, too: Gala-tura and Kur-jara. Although in the rest of the text, these beings are referred to as demons, devils and phantoms, we think that these formed beings are a form of energy or organic robots. (10)

In "The Lost Book of Enki", this is how Ninmah, who looks at the Homo Erectus in front of her, explains the fact that the Anunnaki are made of from the same essence as the humans: "Long time ago, our ancestors in Nibiru must have lived like these." And, Enki actually refers to the seed of life from Nibiru flowing into the World during the celestial collision four billion years ago when he mentions "Those like us have come from Nibiru, and they are reproducing."

The Anunnaki are depicted in writing and figures in thousands of tablets and scripts in the form of a human. And, in some, they are depicted with wings as eagle-men symbolizing their faculty for flying.

When we consider that "Eagle" is the code name of the Moon module which landed on moon in 1969, and at the same time, the nickname for the spacecraft Apollo 11, and that the three astronauts proudly self-named themselves with this title, the Anunnaki's depiction as eagle-men in the figures gains meaning. In fact, our astronauts are each an "Eagle Man", too.

There are numerous tablets with information on this, and all of them have depicted the Anunnaki in the form of humans. Greek, Mayan, Aztec, Incan, Turkish and Celtic mythologies all have a tendency to see their deities in the form of humans. Only in Egypt and Far East things get a little complicated.

In Ancient Egypt, the Anunnaki have been symbolized with some animal heads. For example, Thoth has been symbolized with an ibis head, Seth with a donkey's, Ra with a falcon's and Hathor with a cow's. However, both in the Sumerian tablets and the holy books as well as mythology, they have always existed in the form of humans. There are thousands of tablets which has drawings of the Anunnaki showing them looking like humans, and giving descriptions of them in writing telling that they look like humans. Yet, there is not a single piece of information mentioning that the figurines that are claimed to be Reptilians are gods and goddesses. Nevertheless, there is one thing that is clear; even in narrations like the myth of Adam and Eve and the Epic of Gilgamesh, the snake has been the symbol of the Enki clan, and the bull and the eagle that of the Enlil clan.

Face-to-Face Contact with the Anunnaki

For us, who consider the Torah as a book of history, there is much information about the Anunnaki having the human form. The first one of these is Abraham's meeting with the three Anunnaki.

Book of Genesis 18:2 Abraham looked up and saw three men standing nearby. When he saw them, he hurried from the entrance of his tent to meet them and bowed low to the ground.

Hereby the Anunnaki met by Abraham are no different than Abraham himself in form, and they have been referred to as "men".

In the latter parts, the Torah mentions that one of them is the Lord and the two others are helpers, and tells about the conversations in the form of a dialogue. Once again, the Anunnaki are in the form of humans, according to Prophet Lot:

Book of Genesis 19:1-2 Two angels arrived at Sodom in the evening, and Lot was sitting in the gateway of the city. When he saw them, he got up to meet them and bowed down with his face to the ground. "My lords," he said, "please turn aside to your servant's house. You can wash your feet and spend the night and then go on your way early in the morning."

Lot was very pleased to have met with the Anunnaki, but the public were affected by their beauty and lusted for them. In this narrative, too, we observe that they are in the form of humans. In the next sentence following, we understand that they are in the form of men anyway:

Book of Genesis 19:5 They called to Lot, "Where are the men who came to you tonight? Bring them out to us so that we can have sex with them."

Yet in another section, Prophet Jacob wrestles with a man he meets on the road until the morning, and finally decides that he is God. Hereby, it is persistently emphasized that the God of Israel is a man, and that he looks like a human being:

Book of Genesis 32: 24 So Jacob was left alone, and a man wrestled with him till daybreak.

Book of Genesis 32: 25 When the man saw that he could not overpower him, he touched the socket of Jacob's hip so that his hip was wrenched as he wrestled with the man.

Book of Genesis 32: 26 Then the man said, "Let me go, for it is daybreak." But Jacob replied, "I will not let you go unless you bless me."

Book of Genesis 32: 27 The man asked him, "What is your name?" "Jacob," he answered.

Book of Genesis 32: 28 Then the man said, "Your name will no longer be Jacob, but Israel, because you have struggled with God and with humans and have overcome."

Book of Genesis 32: 29 Jacob said, "Please tell me your name." But he replied, "Why do you ask my name?" Then he blessed him there.

We think that Jacob would not have dared to wrestle had the Anunnaki not been in the form of humans.

We can easily see in the drawings of the palaces of Assyria, Babylonia and Sumer that the Anunnaki, who are depicted tall and with beards, are in the form of humans. However, we should not cast out the probability that they could have appeared in different forms to humans by means of a technological or otherwise method.

According to the columnist Deniz E. Dogru who writes in Sec Haber, the Anunnaki wore costumes suitable for the geographies they visit. This in return gave way to different depictions of them by different human beings. Then again, sometimes, their costumes were so unique that they ended up making them look like the animals around them. While this information was passed on from one generation to another, their actual appearances were forgotten, yet, the animals symbolizing them were not. No matter what, the Anunnaki are in the form of humans in essence. As one can see on a statue in Peru, the beings named as Reptilian are perhaps the Anunnaki in costumes. (Figure 1)

Photo 1: A statue in Peru

About the Discovery of Nibiru/ Planet X, the Planet of the Anunnaki

The Sumerians wrote that the Anunnaki arrived on Earth from their own planet Nibiru, 120 Shars, that is; 432,000 years before the Great Flood.

They believed that Nibiru, which to them was one of the members of the family just like the Sun, the Moon and the other ten members, is the outermost planet of the Solar System. According to what they have written, it takes Nibiru to complete one orbit in 1 shar; 3,600 Earth years. This turning around the Sun takes Nibiru to a "station" at a very far corner of the skies. Then, it brings it back around the Earth passing through Mars and Jupiter. And it is right at this position that the planet's name "Nibiru" is derived, which means "intersection", and turns into a source for the symbol of the cross form.

Based on the tablets, Nibiru of the Anunnaki is the outermost planet of our Solar System, which is known as Planet X. We believe that both NASA and ESA have been monitoring this planet since 1983 when it was discovered by the satellite IRAS. Scientists working at the Kobe University in Japan have pronounced that they have been convinced of the existence of a planet at the outermost edge of the Solar System. Researchers have stated that it is only a matter of time before the announcement of the mysterious "Planet X" based on theoretical calculations using computer simulations. (11)

The existence of this planet have been partially announced by Mike Brown and Konstantin Baytgin, two NASA scientists, in January of 2016. In an article based on the new findings about our Solar System, which was published in the Astronomical Journal, it was emphasized that the new planet might have a distance of 20 to 100 billion kilometers and an orbit of 10 to 20 thousand years. Let us add right away; this orbit in mention has been calculated based on a circle. If Planet X's orbit is calculated based on an elliptic route rather than a circle, then, it will have a 3 to 4-thousand-year span rather than the 20 thousand year orbit as mentioned in the article. All this confirms the existence of the Planet Nibiru of the Anunnaki, which has a 3,600 year orbit, as mentioned by the Sumerians. (12)

Mike Brown, an astronomer who has been a planet astronomy professor at Caltech since 2003 has gone even further, and presented the evidence for this planet on YouTube on March 3, 2017. What is surprising is that Mike Brown, whom we encounter as the friendly smiling face of NASA, carries on his research under the title "The 9th Planet: Planet X". In other words, even if the existence of this planet is accepted, it will be mentioned as "the Sun, the Moon and the nine planets in our system". (13) Yet, based on what the Sumerians tell in Enuma Elish, and the tablet numbered VA 243 which is on display in the Berlin State Museum, our system consists of the Sun, the Moon and ten planets. The only difference between the information we gather from NASA and the Sumerians lies on Pluto, because, based on nothing but an alteration on the description, and a voting following this during a conference in Prague in 2006, it has been dropped from the list of planets. (14)

Once again, we see Mike Brown, the smiling face of NASA's Planet X research, as the chief of the team which dropped Pluto from the list of planets during that period. Mike Brown has even written a book called "How I killed Pluto" in which he talks about this backstage success. (15) We think that it is an attempt for perception management. Pluto has been dropped from the planet list just to be able to say that the Sumerians were wrong; "There aren't ten planets. Only nine." Besides, despite the heavy objections, NASA fulfills its duty with success, too: Mike Brown, who has succeeded in dropping Pluto from the planet list, can be announced as the hero of the discovery of Nibiru/ Planet X.

Nevertheless, Professor Renu Malhotra from the University of Arizona, who is also one of the legendary astronomers of NASA, has given quite a lot of information about the research on the new planet during his explanations at TedX on July 17, 2017. (16) Based on the news of The Sun newspaper, on October 16, 2017, NASA has finally accepted the existence of "The Ninth Planet (Planet 9)" which may be changing the entire destiny of the Solar System. (17)

The Book of Genesis is being written

Based on the information we gather from the ancient texts, the chief of the astronauts, who came from Nibiru to Earth, was the so called Anunnaki, E.A which meant "that whose home is water". After descending to Earth for the ex-king Alalu, and establishing the first Earth station Eridu, E.A gets the name EN.KI meaning "the master of the Earth". According to the texts, when ENKI came to the Earth there was nothing. He landed in the marshes at the entrance of the Persian Gulf. Later on, he told what he did on each Earth day together with the fifty men he brought along with him.

Once on Earth, Enki finds the relation between the Sun and the Earth quite interesting. The first thing he does is to calculate both the distance and the daily period of the Sun. He understands that a day has twenty-four hours, and thus, he distinguishes day and night. He decides that work should be carried out during daytime. He dedicates the next day to figuring out the cycle of the sky, the clouds, condensation and rain. The following day he examines the marshes, orders channels to be built and trenches to be dug in order to dry the marshes and control the waters, and structures to be built using the clay bricks made. He builds a water house with a dock and other facilities to the edge of the marshland after connecting River Tigris and River Euphrates to each other by means of channels. On the sixth day, he considers food supplies; once he sees that fish and birds can be consumed, he relaxes. And the seventh day, he announces it to be a day at rest for his team whom had been tired with all these works. From this point of view, the Book of Genesis from the Torah gains a different meaning.

Once the Anunnaki created the form of life for themselves, the next job for them was to research the gold metal existing in the seas. In fact, there is only one goal for his entire arrival; that is to extract gold from the waters of the Persian Gulf and from the shallow marshlands extending from the Gulf into the inner parts of Mesopotamia. However, when the story is examined into its details, one sees that the plan did not fall into its place. The amount of gold production was much lower than expected, and they needed extra personnel to increase it.

We do not think that they were so keen on finding gold because of their desires for jewelry production or some other artificial use, because for the following period of thousands of years, these visitors were never ever depicted for even once with gold jewelry on them. On the other hand, in Mahabharata, and also in other Hindu texts, there is mention of aircrafts covered in gold. Therefore, we can suggest that the gold was used in space programs, or it served for a necessity on their own planet. Even today, the gold metal has crucial importance for spacecrafts. However, when we consider the intensive search of the Nibiruns for gold on Earth, and the effort they showed in transferring profound amounts of gold to their planet, we can easily concur that this goal served another vital purpose.

The gold metal is not affected by ordinary acids, and can be easily worked on because of its softness and malleability. It is not affected by oxygen and water, because it is a very stable element which is not easily reactive. This heavy metal is one of the most valuable metals in the modern age due to its shiny yellow color and glow, and its characteristics such as no oxidizing, no tarnishing and no corroding. With its unmatched characteristics, this metal really serves a vital need: maintaining the continuity of life on Nibiru is in question. Nibiruns used the gold particles in protecting the ever-thinning atmosphere of Nibiru against a thinning reaching dangerous levels. Today, our scientists are planning to do the same for the weak points of the ozone layer.

In the meantime, there are other arguments about the interesting characteristics of the mono-atomic gold, and if these are true, we can conclude that the search for gold not only serves in fixing the atmosphere, but also in much deeper rooted research. Enki, the son of the Nibiru king, who was appointed for this job, was a bright scientist and engineer who had the title of NU.DIM.MUD. They had planned to get gold from the sea water through laboratory processes. However, as the plan did not go as foreseen, they started with mining research. As a result of the research, they saw that gold was plenty in Africa, particularly in Ab.Zu, which meant "the oldest source" in Africa. Even today, the word "zaab" is being used for gold in the Sami languages.

At this point, this question comes to mind: "Has the Nibiruns' need for gold come to an end today?" In order to give the answer to this question, first we need to ask where the gold being collected in the world is now. Today, among all the nations in the World, only Venezuela has been able to retrieve its gold back from the world banks, and only after a full-court press. (18). And, Germany, being encouraged with this, could only retrieve some part of it. (19) It is said that the gold reserves belonging to all the remaining nations are being kept in the special depots of the banks in overseas countries. Based on the claims of economists such as Kaan Sarıaydın, the amounts of gold in mention do not exist in the depots. (20) If this is true, these questions come to mind:

Is the World's economy system being managed over the gold that "does not exist"? So, where could this gold be? We wonder if some entities out there might have replaced gold with paper, and established a system based on fiduciary money. And thus, might they have created a systematic order for the flow of gold to Nibiru? If we answer these questions with a "Yes", can we not say that the Anunnaki are controlling the political and economic systems of today? Now, let us get to know these Anunnaki a bit closer:

The Deities of the Polytheistic Times: The Anunnaki

The Sumerians believed in the existence of deities from "the skies". The Enuma Elish texts, which talk about the times "before things were created", mention such deities from the skies as Apsu, Tiamat, Anshar and Kishar. There are no claims about the deities in this category to have come out of the Earth. When we look at these deities, which existed before the Earth was created, in more detail, we notice that these are the celestial bodies that form our Solar System. As mentioned in Zecharia Sitchin's books, the alleged Sumerian myths are in fact, definitive and scientifically reasonable cosmological concepts that have to do with the creation of our Solar System.

They have the "Great Deities Pantheon" in command. They are ruled by a Council of Divinities and they have kinship to each other. When the more secondary nieces, cousins, grandchildren and so on are left outside, a much smaller and consistent group of divinities appears. Each divinity has a role to play, each has specific powers or responsibilities.

Besides, there are secondary deities who are "from the Earth". The cult centers for these local divinities are usually towns in the countryside. At most, they take some limited responsibilities just like in the example of the goddess NIN.KASHI ("Lady Beer") who overlooks the preparation of the drinks. There are no heroic stories told about them. They do not have fearsome weapons, and the other deities do not tremble on their orders.

Between the two groups, there are the Sky and Earth deities who are called as the "Ancient Deities". Those are the archaic deities of the myths, and as per the Sumerian belief, they had descended from the skies on to the Earth. They are national rather than local deities. More accurately, they are international deities. Some of them existed and were active on Earth even before the human beings. In fact, the existence of human beings is a result of the purposeful creative initiative of these deities. They are powerful, and they possess capacities beyond the understanding of the mind and mortal capabilities. Nevertheless, these deities not only look like human beings in appearance, but they also both eat and drink like them, and display human feelings such as hatred, loyalty, and betrayal.

Although the roles and hierarchical status of some of the main divinities have changed within thousands of years, a certain deity group has never fallen from its place at the peak, nor has it lost anything from the respect on the national and international level. Looking closer to this central group reveals a divine family picture in which the dynasty of deities is bitterly divided despite their close affinity.

The Anunnaki and the Rank-Frequency Distribution

The status of the children and grandchildren of Anu, Enlil and Enki in the dynasty of ancestry becomes clear through the appointment of numerical ranks to each one. The modern day discovery of this unique Sumerian system also reveals the members of the Grand Council of the Sky and Earth deities. Through this we learn that the Superior Pantheon consists of twelve divinities (The Council of Twelve). The first hint on the application of a coded numerical system for the great deities was reached at through the discovery of the use of number 30, 20 and 15 as a substitute for the names of Sin, Shamash and Ishtar. The highest number 60 in the Sumerian sexagesimal system is designated to Anu, whereas 50 is for Enlil, 40 for Enki, 30 for Sin, 20 for Shamash and 10 for Adad. Number 10 and its multiples in the cardinal number 60 has been designated to male deities. Consequently, the numbers ending with a 5 have pointed out to female deities; 55 for Antu, 45 for Ninlil, 35 for Ninki, 25 for Ningal, 15 for Inanna and 5 for Ninmah.

At this point, it is crucial to remember Nikola Tesla's quote: "If you want to learn the secrets of the universe, think in terms of energy, frequency and vibration." Anunnaki, too, seem to have elucidated energy, frequency and vibration during the later periods, if not at the time of their arrival.

We think that the Anunnaki, who had elucidated the knowledge of frequency, first decreased the frequency levels of humans through junk genes, and therefore, they suddenly became invisible for us. In fact, simply by looking at the flow of history, we can clearly follow the process of visible deities becoming invisible. We read that the deities and the human beings used to live together during the early times of the

Sumerians as well as the mythologies of various nations. The Anunnaki address human beings after the 2nd millennia BCE at the top of the ziggurats in the cult cities through intermediaries (priests) they choose once year. Once we reach the 500s BCE, the deities are never seen, but they can reach at people at any time they like. And then on, they contact with people through dreams, visions and intuitions. The Anunnaki have been given different name in the different cultures of the world. In our society, they are sometimes called as angels or demons. They are never a Creator. This point has to be well distinguished. In many tablets, the Anunnaki have mentioned about their belief in that great power which has created the universe by referring to him as "The Great Father who has created everything". Consequently, the Anunnaki might have inserted the information on frequency, which they had gathered through scientific research and advanced technologies, into our lives by using the concept of "dimension".

There are several other deities in Sumer; these are the children, the grandchildren, the nieces and cousins of the "Great Deities". Separately, there are about a few hundred more deities appointed to several duties. Yet, only twelve of them form the great group, and we think that this great group is renewed at the turn of every era. According to Sitchin, the so-called "Council of the Twelve" consists of the following deities in the Bull Era:

Anu, Antu, Enlil, Ninlil, Enki, Ninki, Ninmah, Ninurta, Sin, Adad, Shamash, and Inanna.

Sources:

1. Çığ, Muazzez İlmiye, Sümerlilerde Tufan, 10. Basım, Kaynak Yayınları, İstanbul, 2015, s. 41
2. Çığ, Muazzez İlmiye, Sümerlilerde Tufan, 10. Basım, Kaynak Yayınları, İstanbul, 2015, s. 41
3. http://beforeitsnews.com/paranormal/2014/02/president-eisenhower-had-3-secret-meetings-w-aliens-in-1954-including-the-anunnaki-former-pentagon-consultant-military-deep-insiders-claim-2464774.html
4. http://enkispeaks.com/2012/06/29/president-eisenhower-sold-us-out-to-ets/
5. https://www.ancient-code.com/the-message-of-the-anunnaki-we-are-here-among-you-now/
6. http://sitchin.com/
7. https://en.wikipedia.org/wiki/Zecharia_Sitchin
8. http://www.independent.co.uk/news/world/middle-east/iraq-spaceships-transport-minister-kazem-finjan-iraqi-sumerians-space-travel-7000-years-ago-a7340966.html
9. http://www.webtekno.com/nasa-uzayda-kullanmak-icin-her-yere-girebilen-minik-robotunu-duyurdu-h28861.html
10. Inana's descent to the nether World
http://etcsl.orinst.ox.ac.uk/cgi-bin/etcsl.cgi?text=t.1.4.1&charenc=j#
11. http://www.thelivingmoon.com/43ancients/02files/Nibiru_Planet_X_001.html
12. https://solarsystem.nasa.gov/planets/planetx/indepth
13. http://www.sitchin.com/
14. https://www.youtube.com/watch?v=v-ktWBtt7sc
15. http://www.wikizero.info/index.php?q=aHR0cHM6Ly9lbi53aWtpcGVkaWEub3JnL3dpa2kvUGx1dG8

16. https://www.amazon.com/How-Killed-Pluto-Why-Coming/dp/0385531109
17. https://www.youtube.com/watch?v=MptrypvBTag
18. https://tr.sputniknews.com/bilim/201710161030609308-nasa-gizemli-dokuzuncu-gezegen/
19. http://www.gazetevatan.com/altinlarini-geri-aliyor-413634-ekonomi/
20. http://www.hurriyet.com.tr/almanya-216-ton-altini-geri-getirdi-40360971
21. https://www.youtube.com/watch?v=3qYAEAz76RQ

SECTION 3

THE ROYAL FAMILY OF THE ANUNNAKI

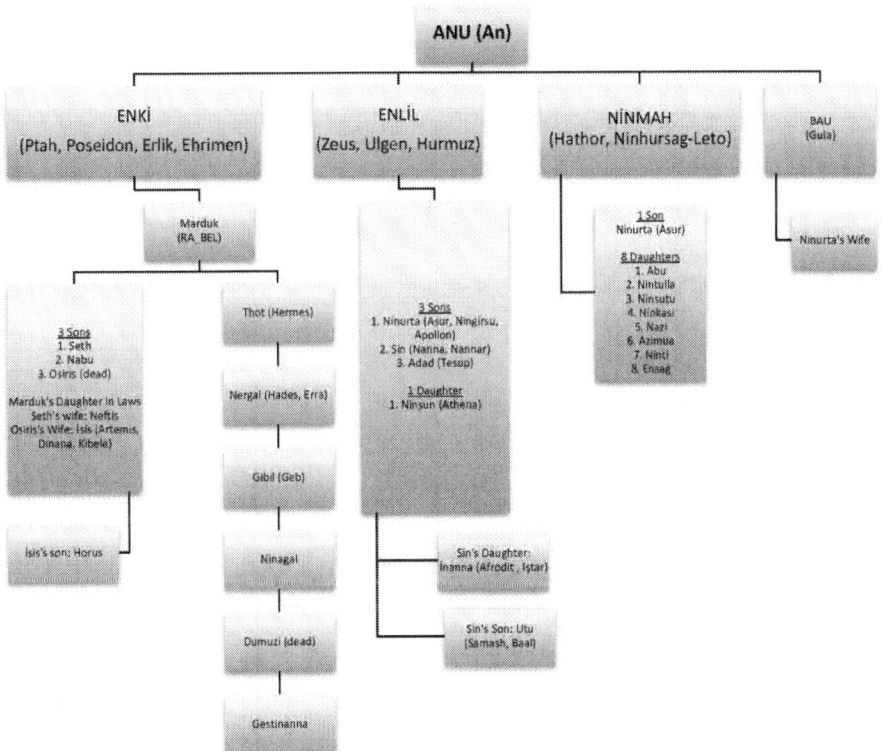

"Let the future tell the truth, and evaluate each one according to his work and accomplishments. The present is theirs; the future,

Nikola Tesla

The Hierarchial Order with the Anunnaki

An/Anu, the Nibirun King, and the greatest deity known is the owner of everything, and the Anunnaki are ruled by the Anu Clan. All ceremonies are held in his name, all meetings begin with his name, and all resolutions adopted in the assembly are put into action within his jurisdiction. On top of this, both on Nibiru and Earth, there is a monarchy-like democratic order which has a constitution. In the ancient tablets, Enlil says that his powers not only come from the above, meaning Anu and Nibiru, but also from below, from "The Judging Seven, The Council of the Twelve and The Great Deities Assembly".

Unluckily, we do not have much information about how the assembly on Nibiru runs. However, as the implementer of Anu on Earth, first we come across with the "Council of the Twelve". This is a world council consisting of twelve people with the executive powers. The discussions while taking the most crucial and providential decisions here sometimes turn into long debates and frequently into a heated battle of words.

The leader of this assembly is chosen from the Enki and Enlil clans respectively every 2160 year precession round of the constellation, and the membership is distributed equally among the two clans. The Council of the Twelve is the highest rank any Anunnaki can reach on Earth. In the Greek Pantheon, we see gods and goddesses who go in and come out of this assembly quite frequently, and we observe that the members act in somewhat in the capacity of the government. In every mythology in the World, we come across with a similar high council.

We can clearly see the names of the members of the Council of the Twelve, which was led by Enlil in the Bull Era. On the other hand, the names of the members during the Ram, Pisces and Aquarius are not clear, and one can only suggest guesses.

The leader of the Aquarius Era we are just entering, and which will last for 2160 years, is Enki on whom the Aquarius

Constellation is also designated. Yet, the other eleven people to sit next to Enki is open to arguments. Our guesswork points to Damkina, Marduk, Thoth, Isis, Hathor, Enlil, Ninlil, Ninurta, Bau, Shamash and Inanna as the other Anunnaki members whose names we will hear frequently in the new era.

Besides the Council of Twelve holding the power of execution, there is another council with seven members holding the power of judiciary. The members of this council referred to as "The Judging Seven" in the tablets varies constantly. The expulsion penalty inflicted on Enlil is discussed by the greatest fifty deities first, and then, the Judging Seven decides on imposing the penalty. In such cases of Alalu and Enlil getting the expulsion penalty, and Anzu getting the capital punishment, we always see the Judging Seven. Furthermore, we can conclude that, in the Anunnaki society where even Enlil is imposed to expulsion penalty, the laws are strict, and they do not extend any prerogatives to anyone.

We also learn about the existence of the Council of Forty and Council of Fifty from the tablets. The Council of Twelve constantly has to account for their actions in these councils. The Great Deities Council, which consists of fifty people, are called as "Ilani rabuti sha mushimu shimati" in the Acadian texts, meaning "The Great Deities who determine the Fates". We see that the upper level Anunnaki and the secondary class royal family members such as the children, and nieces, operate in here in a structure similar to the TGNA (Turkish Grand National Assembly). The members of the Council of Twelve are always chosen from the Council of the Forty and the Council of the Fifty, and it also works reversely; those who come out of the Council go back to these assemblies.

The Anunnaki, who gathered in Hattusha, Yazilikaya in 2000 BCE under the leadership of Anu, have accepted the matter of leaders in the precessions, right after a holy matrimonial ceremony. Following the rejections lasting for about a hundred years, they gathered in Babylon in 1890 BCE to sign the agreement. The Enuma Elish Epic, which tells how the old deities gathered in order to announce the superiority of Marduk, describes the get-together of friends who had not seen each other for a long time. (1) According to the epic, Marduk's superiority has been tested and confirmed in the Assembly of the Anunnaki.

This ancient text shows that the Assembly of the Fifty and the Judging Seven had separate meetings, and that the official announcement of the leadership was done by Enlil after consulting to Anu or after getting his approval. The eminent Babylonian king Hammurabi praises the superiority of his deity Marduk in the preface of the famous code of laws inscription with the following wording:

"When the lofty Anu, master of the gods who descended from the skies on to Earth,

and Enlil, lord of heaven and earth, he who determines the destiny of the land;

committed the rule of all mankind to Marduk, the chief son of Enki;"

The Nibirun King: An/Anu

Although his name does not appear often in the Sumerian and Akkadian inscriptions, the greatest deity is Anu. He is also known as An. The official wife of Anu is Antu. On the other hand, he has numerous children from various concubines. Anu has fourteen children born from his official wife Antu. Separately, he has six concubines. In the tablets, we come across with two sons and two daughters of Anu who are related to Earth. His sons are Enlil and Enki, and the daughters are Bau and

Ninmah, otherwise known as Ninhursag. As Enlil was the son of Anu and Antu, kingship was entitled to him.

Enlil, who is the second of the two sons of Anu, has been acknowledged by the Nibiru Assembly as his successor, because he was born from Antu, who was the official wife of Anu. Consequently, although Enki is the first-born, the clashes between Enlil and Enki, who is his step brother born from a concubine, are mentioned in all mythologies.

In the atmosphere of Nibiru, there was trouble similar to the hole in the ozone layer of ours, and it was threatening the future of the planet. This problem was resolved by means of the gold mineral which existed in ample amount on Earth. The agenda consisted of establishing a colony on Earth in order to mine the gold mineral. Therefore, Anu traveled to Earth around 400 thousand BCE For distribution of duties, the dice was rolled. The kingship of Nibiru fell on to him while Enlil was entitled to the Kingdom of Earth, and Enki's jurisdiction covered the seas and the mines. Anu became well-known as the Great Deity after that day. For the most important matters of Earth, he was the one who was always consulted. He was always entertained at top levels during his occasional visits to Earth. 3760 BCE, which was the year he descended on to Earth during the Sumerian times, has been recognized as the starting point of the Nippur calendar.

The leading deity among the celestial and earthly deities is Anu. He is the grandfather of the deities, the king of the deities. The abode of Anu, and his kingdom's center is in the Skies, in other words; on Nibiru. Nibiru is the outermost planet where the rest of the Celestial and Earth deities went to or sent messages to whenever the Anunnaki needed personal advice or favors, whenever they gathered in order to resolve a matter of dispute among themselves, or whenever they were on the brink of important decisions. The Sumerian texts tell that not only the other deities, but, sometimes, the chosen mortals, too, were permitted to go to Anu's abode, many times for the purpose of getting rid of mortality. One such story is about Adapa. Adapa is so exceptional, and so loyal to the deity who created him that Ea arranges a visit to Anu for him.

Under this guiding Adapa "moves towards the skies. He ascends to the Skies, and approaches the gate of Anu." However, when he is presented the gift of mortality, Adapa refuses to eat "The Bread of Life". He thinks that the raged Anu presented him poisonous food. Consequently, he returns to Earth as an anointed priest, but, still a mortal. The Sumerian claim that not only the deities, but also the chosen mortals can ascend up to the Divine Adobe in the skies, echo in the stories of the Old Testament talking about the ascend of Enoch and Prophet Elijah to the skies.

As Anu did not have a permanent duty on Earth, he was not ordained with preeminence in his own city or cult center. However, in Uruk, which was under Goddess Inanna's jurisdiction, a building called as the "the high house" was erected in Anu's honor. Our suggestion is that Anu gives this building as a present to Inanna after his visit to Earth.

Anu is not only the Great King, and the God of Kings; based on several texts, others can become the king only through his grace. According to the Sumerian tradition, sovereignty flows only from Anu, and the corresponding title for kingship is "Anutu" meaning "Anu-ship".

Anu's symbols are the divine horned crown, the scepter as the symbol of might, and the shepherd's crook for protection by the shepherds. However, the shepherds' crook is nowadays mostly in the hands of bishops rather than kings, but the crown and scepter are still being held by the kings the human beings have left to survive in a few thrones.

King of the Earth: Enlil

Enlil is the most frequently mentioned deity in the Sumerian and Akkadian inscriptions. As he has chosen the eagle as a symbol, his whole clan has used the double-headed eagle as a symbol, too. Although he was known to be a soldier and an executive, he is the son of Anu, the Great King of Nibiru, and the most powerful deity on Earth. We see him as the greatest deity together with his brother Enki in every mythology. The Greeks recognize him as Zeus, the Romans as Jupiter, the Turks as Ulgen, the Persians as Ahura Mazda, the Iranians as Hormuz, and the Americans as Viracocha. When there was talk of establishing a colony on Earth for the gold mines, Enlil descends to Earth on duty around 400 thousand BCE Since the day when the leadership of the Earth has fallen on him, he is known to be the "Lord of the Commands".

During the early times on Earth, Enlil founds five cities in Mesopotamia while building a house for himself in the coolest part of the cedar forests in the city of Baalbek. And in the city of Nippur, he establishes a DUR-AN-KI point; in other words, the "Bond Sky-Earth". He coordinates the flights between Nibiru and Earth from here, and also keeps the Tablets of Destinies here, too.

Before descending to Earth, and as a single prince, Enlil was in a love relationship with his step-sister Ninmah, the scientist and healing princess of Nibiru. Ninurtu was born as a fruit of this love affair. The great king Anu had once wished for Ninmah to marry Enki. However, when Ninmah fell in love with Enlil and stood up to him, he abolished marriage for her.

According to what the tablets tell, Enlil sees a young girl having a bath by the river while strolling around, and he gets struck by her beauty. He invites her to his home. After spending some time with her, he wants to have intercourse with her, but the girl resists. Eventually, Enlil rapes the girl. When this rape comes to the knowledge of others, Enlil is brought to the presence of the Judging Seven, and he gets expulsion punishment. However, once he agrees to marry the girl, his is granted a pardon. The girl name changes to Ninlil, meaning the wife of Enlil. (2) Two sons named Sin (Nannar) and Adad (Tesup) and numerous daughters are born out of the marriage of Enlil and Ninlil. Ninlil's corresponding figure in the Greek mythology is Hera. Contrary to what is being told in the Greek mythology, Enlil adopts monogamy after getting married to Ninlil. He leads a scrupulous and virtuous life, and he stands out in all discussions with his conservative opinions. (3)

The Mother of Life:Ninmah

Ninmah is one of the goddesses whose name we come across with most frequently in the Sumerian and Akkadian inscriptions. The goddess, who is known as the scientist and healing princess of Nibiru, is the daughter of the Great King Anu. Since she is one of the most powerful goddesses, she corresponds to the mother goddess almost in every mythology. She is known as Hathor among the Egyptians, Ninhursag and Ninti among the Akkadians, and Leto among the Greeks. In reference to the struggle between Enki and Enlil, Ninmah is sometimes on Enki's side, and at other times, on Enlil's. However, in general, her neutral approach is recognized by all the Anunnaki leaders. Consequently, she is deemed as the most respected goddess.

When the necessity of establishing a colony on Earth for operating the gold mines was on the agenda, Ninmah comes to Earth on duty. Enlil allocates Shurubak, one of the five cities he had established in Mesopotamia during his early times on Earth, to Ninmah. Shurubak turns into a full-fledged city of healing. Later on, a laboratory is set in the cedar forest in the city of Baalbek for Ninmah and her healing team.

Based on the Sumerian texts, Ninhursag, who followed the formulas Enki had invented, intervened on humans. (4) We think that by "intervention" they refer to the intervention applied on Homo Erectus around 300 thousand BCE Following consecutive technological and scientific processes, the existence of the first humans occur. In the capacity as the chief healer and the figure responsible for the medical processes in the healing team, Ninmah has been called as Goddess NIN.TI ("Lady Life") for this role she played.

In the ancient texts, Ninmah/ Ninhursag has been referred to as the "Mother Goddess" since she was the one in charge of giving life to the deities and the humans. Her nickname is "Mammu", and this word is somewhere between the English words "mom" and "mama" which means mother. Her symbol is the "Blade", that is; the blade the midwives used to cut the umbilical cord with right after birth.

In many ancient cities, we observe her footprints. Besides Baalbek and Shurubak, today, in Egypt's Dandera city, there is a temple dedicated to "Hathor", which is the Egyptian version corresponding to her. Other than this temple, where the famous Egyptian light bulbs are depicted on the reliefs, we think that Ninmah's footsteps can also be found in the ancient cities of Gobeklitepe and Rhodiapolis in Turkey.

The Leader of the Age of the Aquarius: Enki

In the Sumerian and Akkadian texts, the most frequently mentioned God, only in a close second to Enlil, is Enki. As Enki had chosen the double helix serpent symbol standing for the DNA, his whole clan had used the symbol of serpent, too. Enki, who is known to be a great scientist and a genius, is the son of Anu, the Great King of Nibiru and one of his concubines. He is the second most powerful deity on Earth. In all the mythologies, we see him as the greatest deity next to his brother Enlil. He is known as EA among the Akkadians, Poseidon or Serapis among the Greeks, Erlik among the Turks, Ptah among the Egyptians, and Ahriman among the Iranians. Enki descends to Earth together with the group of fifty astronauts around 443 thousand BCE. We can consider him as the first figure to step on Earth. Because he has arrived in the sea, he has found a place for himself in mythology as the fish man. Many times he is depicted together with Ninmah in the laboratory environment. In the sources of the Enlil clan, he has always been depicted as the devil.

Enki builds his big house in the so-called city of Eridu which he had founded right on the edge of the marshlands. He builds sailing ships in the ports where precious metals and semi-precious Stones are brought into, because he is the lord of the seas and oceans. Simultaneously, he has the sovereignty over the Abzu mines, in other words, the underworld. Although Enki is married to Ninki, he has several children from many different concubines and affairs. The tablets give wide coverage on Enki's love affairs. There is even a book called "The Myths of Enki, the Crafty God", written by Samuel Noah Kramer, who is recognized as one of the leading Sumerologists in the world. The book talks about nothing but the craftiness of Enki. Enki has six sons that are known. The first of these sons is Marduk who was born from his wife Ninki. The other sons are Thoth, Dumuzi, Nergal, Ninangal and Gibil. Thoth is born out of the love affair of Enki and Ereshkigal. Dumuzi on the other hand, is born from Enki and Ninsun.

In the capacity of the chief scientist of the deities, Enki is the one who determined how the creation of the humankind would be. We see him in the holy books through the Adam and Eve story as the serpent which deceives Eve to eat the apple, and in the Epic of Gilgamesh, as the serpent which steals the plant of immortality from Gilgamesh.

Enki has been described as a humanistic figure in the tablets. He has been by the humans through thick and thin, and has always supported the humans in the discussions in the Assembly of the Anunnaki. Based on the Akkadian texts on the Flood circa 11 thousand BCE, it is Enki who saved Utnapishtim during the Great Flood, and who protected the animals and the seeds of the plants by putting them in the ship he ordered to be built. His Egyptian version Ptah has been described as the God who has given life to Egypt. Similarly, it is Enki who advised the humans to escape to the mountains or to enter the underground shelters when the radiation loaded clouds of the nuclear weapons used over the Sinai Peninsula and the Dead Sea threatened the Sumerian cities in 2023 BCE. Recognized as the leader of the

Aquarius Era, Enki and his scientific identity has played a major role in the technological and scientific developments of the last five centuries.

A Healer and a Physician: Bau

Bau is the healing princess and scientist of Nibiru just like Ninmah. She is the younger daughter of the great King Anu, and the wife of Ninurta who is the legal inheritor of Enlil. She is also given the title Gula, meaning "Great". As far as we can follow from the texts of King Gudea, she lived in Lagash, which was her cult center, together with her husband Ninurta. There is even mention of her personally being in Girsu, the holy section of the city, circa 2100 BCE. The marriage of Ninnurta and Bau has never been subjected to any kind of infidelity. Expressions such as the "Child of the Holy Moment", "Lady", "elegant woman", "Enlil's darling who has been adorned with attraction", "great resentments" and "the one coming out of heaven" have been used in the texts for Bau. Bau has a major role in Gudea's election as the king of Lagash. (5)

The elegies referencing the sudden destruction of the Sumerian cities imply to a nuclear attack. We think that this nuclear disaster occurred in 2023 BCE. The nuclear cloud, which formed as the consequence of the disaster, sweeps the Sumerian and Akkadian lands while being drifted to Mesopotamia. The "Lamentation Texts" that describe this catastrophe and destruction mention that Goddess Bau was all by herself in Lagash. Bau chooses not to abandon her people, and stays in the region while waiting for Ninurta to return. While Bau keeps on lamenting, the nuclear cloud reaches Lagash, and Bau's hesitation in leaving the city almost costs her life. She is saved on the last moment. However, as she is affected by the radiation, she goes through treatment.

Sources:

1. Enuma Elish http://www.piney.com/Enuma.html
2. Enlil and Sud http://etcsl.orinst.ox.ac.uk/cgi-bin/etcsl.cgi?text=t.1.2.2&charenc=j#
3. Enlil and Ninlil http://etcsl.orinst.ox.ac.uk/cgi-bin/etcsl.cgi?text=t.1.2.1&charenc=j#
4. Enki and Ninmah http://etcsl.orinst.ox.ac.uk/cgi-bin/etcsl.cgi?text=t.1.1.2&charenc=j#
5. A tigi to Bau for Gudea http://etcsl.orinst.ox.ac.uk/cgibin/etcsl.cgi?text=t.2.3.2&charenc=j#

SECTION 4

THE ANUNNAKI BASES IN THE SKIES

"One day human beings will travel in space.

When they stop by in Mars, they will see some traces and signs I have left for them."

Muhyiddin Ibn Arabi

The Anunnaki are on our Earth for approximately the last 450 thousand years or so, and we can follow their traces through the written works, inscriptions and megalithic structures they have left behind. Logic dictates that this species, which we think have arrived from the outermost planet of the Solar System, must have established bases to serve as stations on the planets or satellites between the Earth and Nibiru. In this section, we will be analyzing the path the Anunnaki followed, the bases they established, and the works they have left behind. Additionally, we are going to research into the course of the human beings' launch into space, and the purpose of the colonies they want to establish on the other planets or satellites.

Is the secret to longevity on Mars?

217 years before now, a scientific report prepared by the chemist and physicist Sir William H. Wollaston was read at the Royal Society in U.K. This report on the lighting of London at night got a standing ovation. In this report, Wollaston was suggesting that a slice of the Moon be cut and brought to Earth for use of lighting at night. (1) This interesting goal of Wollaston did not come true, but just a mere 169 years after, humanity landed on the Moon. In this journey which began with dreams, humans nowadays are planning for manned flights to Mars.

The Dutch entrepreneur Bas Lansdorp has initiated the Mars One Project in order to establish a human colony on Mars on 2025. The Mars One Project, which was announced in June 2012, consists of a few phases. The program plans to send a communications satellite and an information apparatus until 2024, and then, in 2027, four astronauts to establish a permanent habitat. The next step is to send groups of four people biennially to Mars. The Project will run with a total of 24 people. Yet, 200 thousand people have applied for it. It is estimated that the colonization will start between 2025-2027 (2)

Space X Corporation, which was founded by Elon Musk, who happens to be one of the founders of PayPal and the CEO of Tesla Motors, aims to lower the cost of space transportation in order to make the colonization of Mars come true. Elon Musk is seriously calculating how to take people to Mars, and how to form a city in the next fifty years where a million people will survive on their own. In this respect, Space X stands as the first private sector corporation to launch a spacecraft into the space (December 9, 2010), to put it into orbit, and to bring it back (Dragon) successfully to Earth. It is also recorded as the first private sector company to have sent a spacecraft (Dragon) to International Space Station (ISS) on May 25, 2012. (3)

What is it that makes Mars so popular lately? Why are private corporations besides space institutions such as NASA and ESA so interested in Mars? Sure enough, Mars will be a good stepping stone for the humankind who is making the calculations for expanding into the space. However, we know it for a fact that the private companies in the capitalist system consider things only with respect to investment. So, what makes Mars so inviting in terms of investment? We may think of lucrative future returns of tourism in Mars. There must be people out there who would not hesitate to spend several hundred-thousand dollars simply for curiosity and excitement. However, for such projects in which billions of dollars are invested, at least a thirty year period has to pass before they can earn the amounts to meet the costs, let alone making any profits. For the 24 people colony planned to be established in 2025, a ridiculous 38 dollars was requested from each of the applicants, and there were 200 thousand applications. (4) When we consider this low fee, it is only a drop in the bucket in comparison to the cost, and it makes one wonder: "Who are the real sponsors of this Project, and what is their goal in hiding themselves?" For us, the answer lies, once again, with the Sumerians.

The Sumerians tell us that their holy masters, the "Anunnaki", who have a very long lifespan, and who are taller than us, come from Nibiru. Several tablets such as Enki's Journey to Nibiru (5), Nannar-Sin's Journey to Nibiru (6), and Ninurta's Return from Nibiru (7) tell the journey of the Anunnaki to Nibiru. The Sumerians base the long lifespans of the Anunnaki to this planet. If we are to explain briefly; the Earth rotates around the Sun in 365 days and 6 hours whereas Nibiru rotates around the Sun in 3600 years. The aging process for the humans that takes 365 days corresponds to an expanded 3600 years for those on Nibiru. This means that when the Anunnaki age one year, someone on Earth ages 3600 years. For the Sumerians, who call each 3600 year rotation of Nibiru around the Sun one "shar", the Anunnaki arrived on Earth 120 shars earlier than the Great Flood. For the Anunnaki whose lifespans started to shrink after their arrival on Earth, physicians were brought over from Nibiru, and by way of producing medicine and elixirs from fruits and vegetables from Nibiru, they had stopped the shortening process. Although there is mention of them succeeding, the Anunnaki who arrive on Earth have to return to Nibiru after a short while, and new arrivals by other Anunnaki have to follow. Yet, it is obvious that other measures were taken for those who stayed on Earth. Nevertheless, for the time being, we do not have information on what these measures might have been. This long lifespan of the Anunnaki meant immortality for us. Since 3760 BCE, when people started recording history based on the Nippur calendar, humanity has been continuously searching for this immortality. This adventure, which began with Gilgamesh, is still ongoing. To reach this goal, the limitations of science are being challenged; huge investments are made for operations like head transplantations (8), crazy experiments such as consciousness transfer are being carried out in top secrecy, all kinds of projects aiming to extend the lifespan are being supported, and the results are followed through meticulously. The purpose of this all is the billionaires' wish for a longer life rather than service to science.

It is thought-provoking that the special food and drinks, which were once brought from Nibiru, were grown and created in the laboratory environment and consumed by only the deities, and that these food items were prohibited to human beings. Later on, for the human beings, who inherited the heritage of the Anunnaki, the idea that immortality is hidden in certain food and drinks was observed in various legends.

Herodotus mentions a special place called "Sun's Table" which the Persian King Cambyses searched for in Ethiopia. The king, who thought that this place lengthened the human life, searched for it for a fair amount of time, yet, did not succeed in finding it. Alexander, who came to the region after about a century, also kept on looking for the spring of immortality.

In the later years, as a subject matter for research for the Christian world, the "Spring of Life" has turned into a widespread search even mentioned in the books written by the Europeans going to South America. Columbus and his men embarked on the search for the legendary spring, the water of which "makes the old young again", as soon as they set foot on the land, which they thought to be the islands off the coast of India. The captured "natives" were interrogated, and even tortured by the Spanish so that they would tell the secret about the location of the spring.

The holy scripts, the paganistic beliefs, and the documented stories of eminent explorers, all actually confirm the place which can grant immortality by making the individual forever young by means of its waters and fruit nectars. The Celtic myths mention about Goddess Idunn, who lives in a holy river, and who hides magical apples in a chest. When the deities get older, they come to eat her apples, and they become younger. In the search for immortality, Gilgamesh also returns with the plant of youth after his meeting with Noah. Yet, he loses it to the snake en route.

There is also something called Ab-i Hayat (Water of Life) which is frequently mentioned. Based on several myths, it is only found in the country of the dark. Yet, some sources mention that it is a plant (gillyweed) extracted from the depths of the Red Sea. In the Iranian mythology, it is believed that there is a dark elixir made of blood by bringing together four holy creatures which are Shah-i Galsam, Shahmaran, Imparan, and Simurgh.

In the Turkish myths, Bengi Su (Bengi Water) gives immortality and youth to whoever drinks it. It is extracted from the roots of the Life of Tree. It flows like a river or a creek. Sometimes it appears in the form of foam. For instance, in the Koroglu Epic, it flows as three bodies of spume in one river. It metaphorically represents wisdom, goodness, and leaving behind permanent works. It can even make the dead rise. It surges from a pit under the Ulug Kayin (The Grand Beech). There stands a guarding soul by it. It gives power and strength to whoever drinks it. It cures the sick. Besides, there is folkloric dance called Bengi. It is also thought that Mani, the name of the founder of the Manichaeism religion, derives from Mengu. It is interesting that there is a common belief, which, at many times, comes to being and develops independently from each other, in reference to such a water of life in several different civilizations. (9)

Is the human kind's search for the spring of life in order to postpone death, which has been going on for thousands of years, really meaningless, or will it be successful in reaching this goal as the science and technology advances, and the genetic structure is deciphered? Well, the Gilgameshes and the Alexanders of the current times, who are looking for the answer to this very question, are allocating some of the riches in their hands to the research being carried out on this subject. What we are saying hereby may sound very utopian for the general crowd, or the common people, but the billionaires' notion of leading a long life is one of the driving forces of the Mars projects. We have already mentioned that the Anunnaki age at the turn of every 3600 years, and thus, they seem to the humans beings immortal. The human kind ages in 365 days and 6 hours on Earth. It is estimated that the length of the human life has to do with variables such as the duration of the rotation of the Earth around the Sun, the distance of Earth to the Sun, its orbit, gravity, magnetism, soil, air, water, and elements.

Right. So, what will happen when the human beings go to Mars? Will they be able to adjust when they are there? If they do adjust, will the human beings' bodies change when they begin drinking Mars' water, and eating the plants grown in the soil of Mars? Considering that Mars completes one rotation around the Sun in 687 days, will one age of the human being correspond to 687 days? In short, will someone living on Mars live twice as long as someone living on Earth? Will it be the 150 earth-year old human beings strolling around Mars? To find the answers to all these questions, 100 people with very different metabolisms have been chosen for Mars One Project. This number will drop down to 24, and their developments, and morphoses will be monitored.

According to Jesco von Puttkamer, the NASA Mars Project Coordinator, the human lifespan will be longer because of the Red Planet's gravity at one third ratio of the Earth's. The shortfall of gravity will lengthen the life of the heart, and the digestive system. Consequently, the body will keep its vitality for a longer time. (10) According to Elon Musk, space travel is the best option in order to lengthen the lifespan of the human kind in the universe. Elon Musk's words as "If Space X can continue without me, I would like to go to Mars. I meant it when I said I want to die on Mars." supports what we have said anyway. (11) According to Prof. Dr. Scott E. Solomon, who carries out research on evolution at the Rice University in U.S., the astronauts, who will be forming the colony on Mars, will evolve. (12)

In the future, once the lifespan of those who settle on Mars and adopt to it double, the human beings will aim for new settlements even further than the Sun. Following the footsteps of the Anunnaki, humanity will focus on the natural satellites of Jupiter and Saturn. NASA believes that life may be sustained on as a result of lengthened lifespan.

The second step after Mars will be Jupiter. As a gas giant, Jupiter emits huge amounts of radiation and heat. Consequently, it has a thick atmosphere where violent storms occur. It has been determined that this atmosphere has hydrogen, helium, methane, ammonia, water vapor and droplets, but it is yet to be penetrated. Jupiter, which is at a distance of approximately 800 million kilometers from the Sun, completes one rotation around the Sun in about twelve years. Vortexing storms, and striking lightning and thunders can easily be observed on the surface of Jupiter. Some of the hundreds of natural satellites of the planet rotate very slowly while others rotate very fast.

Although it is not possible to survive on Jupiter, it may be probable on its satellites such as Europa. Giant lava erupting from colossal volcanos, and vein-like crossing patterns on Europa create very interesting images. It is estimated that a big ocean lies under the ice layer that is below the cracked surface of Europa, and that it is protected from freezing by the radioactive dissolution and the friction of the tidal waves. It is also projected that warm, liquid water oases, which would support living organisms, exist once again, under the ice layer. It has been revealed that oxygen is being produced ten times more than hydrogen on Jupiter's satellite Europa, just like on Earth. Researchers have discovered that the radiation scattered from Jupiter separates the frozen waters on the surface of Europa into its molecules, and that oxygen is put out towards the ocean. NASA is planning to send a spacecraft to Europa in 2020s in order to make closer research, and take high resolution photographs. (13)

Since the duration of Jupiter's one rotation around the Sun is twelve years, we wonder whether one age of the people living in the colony to be established on Jupiter's satellite Europa would correspond to our ten ages. While the standard lifespan of humans is seventy on Earth, is it going to be seven hundred on Europa? We guess we will be waiting a little more to get the answers to these questions. Yet, if we assume that it is correct, the next step will definitely be Saturn. Today, we do not have much information on this planet. However, scientists have announced that they have found evidence that there may be life on Titan, the biggest satellite of Saturn. Traces of some simple life forms have been detected to respire the atmosphere of the planet, and to be fed on the materials on its surface.

The scientists, who have analyzed the data sent from NASA's satellite Cassini, told that the surface of Titan is covered with numerous mountains, lakes and rivers, and that Titan is the most similar planet in the Sun System to Earth. (14)

The NASA scientists, who have announced the latest findings on the satellites, which are known as the "Ocean Worlds", have said, especially for Enceladus, which is another satellite of Saturn, "This is the closest we've come, so far, to identifying a place with some of the ingredients needed for a habitable environment," Through the use of the giant spectrometer, it has been observed that the high amount of hydrogen molecules, which have been determined by the Cassini spacecraft, have risen from the icy surface of Enceladus. (15)

Right then. If we assume that a colony is established on Titan or Enceladus, will a human being living there become one year older every twenty-nine years, which corresponds to one rotation of Saturn around the Sun? Will the average lifespan of the people living there be two thousand years?

With all of the above in mind, we conclude that the 3600 year orbit of Nibiru has made the average lifespan of the Anunnaki much longer than ours, and it served for them to become immortal deities in our terms. Now, perhaps, we will be following the footsteps of the Anunnaki, and ultimately, establishing colonies or bases on all these planets and satellites, if not in the near future. In the journey of the human being from the Earth into the space, it is technically and logically correct that Mars will be the first planet. En route to the other worlds, there is need for road-stations due to the celestial motion laws, limitations on weight and energy, and the physical and mental endurance limits of humanity.

A manned spacecraft to go to Saturn leaving from Earth must go to Mars first, fill up fuel and food stocks, and continue the journey. Perhaps, it should also stop by the station to be established on Jupiter's satellite Europa, and complete whatever is in deficiency, and then, continue onwards with the journey once again. If we can envision this, sure enough, the Anunnaki, who are far more advanced in comparison to us, must have thought of it. We are talking about a civilization which could make space travel come true 445 thousand years ago. We are sure that a civilization, who could travel in space in such ancient times, must be traveling in space at much more ease nowadays using more advanced technologies we are yet to learn about. However, when we check into the ancient tablets, we come across with traces showing us that they were still using the rocket technology up until about five thousand years ago. The fact that they were using the technology, which we are using nowadays, up until recently, might be the indication that we are following the same routes. If so, then, they must have established such bases between Nibiru and Earth.

If these bases exist, and if they are still in use, we can suggest that they are built underground in order to be protected from the radiation, the cold and the meteors of the space.

What's happening on the Moon?

The spacecraft named "Lunar Crater Observation and Sensing Satellite" (LCROSS), which was launched by NASA to arrive on the Moon with the purpose of finding out whether there is water or ice on the Moon, divided into two, and crashed on the Moon in a couple of consecutive minutes on October 9, 2009. This was perhaps the most crucial, and simultaneously the most equivocal step taken since 1969.

It is questionable that, just a short while after recording the images of the wreckage and the material for four minutes, and sending the collected data to Earth, LCROSS creates a second explosion by banging on the Moon. Was it really sent for the mission of finding water as it was announced? It had already been determined during the earlier research that there is water on the moon. So, what did NASA aim for with this "bombing"? Although it was known fact that the vacuum of the space would evaporate and diminish the water as soon as it came out as a result of the heat created by a two-ton bomb, why did they insist on the announcement that they were looking for water?

The mind-boggling question here is: What was the real job aimed for LCROSS, which recorded images for four minutes after the rocket Centaur exploded, and then which exploded itself? Why was it more important to observe whether the rocket hit the target area or not, rather than observing the wreckage created right after the explosion? On what basis the impact area was chosen? We wonder if there were things in there which were not desired to be made public. (16)

It is assumed that the Moon is at its pristine condition. To have a look at the Moon is equivalent to having a look at the Creation, because there is no wind, atmosphere, water and any other erosive forces. Humanity has been looking at the Moon for ages, first with bare eyes, and then, with equipment, but only within this century it has been possible to examine it closer. Several Soviet and American unmanned spacecrafts have examined and photographed the Moon by either entering the orbit around it or by landing on it between the years 1959 and 1969.

The Moon is the only celestial body on which human beings have landed and walked on. The first artificial object to go to space by escaping from the gravity of the Earth, and to pass close by the Moon is the Luna 1 satellite of the Soviet Union. The first man-made object to have impact on the Moon's surface is Luna 2. And the first photographs of the other side of the Moon, which normally is not seen, were taken by Luna 3. All of these three satellites were launched in 1959. Luna 9 was the first spacecraft to succeed in soft-landing on the Moon, and the first unmanned spacecraft to enter the orbit of the Moon was Luna 10. Both of these two satellites were launched in 1966. The U.S.'s Apollo program still stands as the one and only space program to have successfully managed manned missions through six landings between 1969 and 1972. Unfortunately, the examining of the Moon directly by people have been terminated since the Apollo Program came to an end. (17)

The latest flights of the Apollo Program had become more scientific. The test equipment and the experiments had been improved. The determination of the landing points had become more scientific. The areas being examined had expanded by the use of surface vehicles. The duration spent on the Moon had increased from hours to days. Even the structure of the crew had changed; a geologist called Harrison Schmitt had been included as a crew member in the last flight.

Since December 14, 1972 when Eugene Cernan walked on the surface of the Moon as part of mission Apollo 17, no other human being has ever walked on the Moon. From the mid-1960s to mid-1970s, there were sixty-five different flight missions reaching the surface of the Moon. The last of these was Luna 24 in 1976. (18)

Some experts claimed that the images of landing on the Moon are fake. However, one can easily see the spacecrafts that had landed on the Moon, and their traces in the numerous high resolution photographs taken by the spacecraft LROC that was launched in 2009. (19) There were also the photographs of the Apollo flags still standing on the surface of the Moon published in 2012. (20) Today, USA Europe, Russia, China, Japan and India are sending their own spacecrafts around the Moon, and carrying out their own researches.

Equipment has been left on the surface of the Moon for long periods of time in order to measure and record the phenomena. Soil samples have been gathered once the surface of the Moon was dug out. Due to budgetary concerns, the data transmission from all the equipment left on the Moon to Earth has been terminated on September 30, 1977. Only the Moon laser distance measurement equipment have been in use until recently, because it is a passive equipment. Out of all this research, the most valuable and rewarding outcome as far as science is concerned, was the Lunar rocks and soil adding up to about 381 kilograms, which was brought to Earth. The examination of these lunar rocks and soil went on for a lengthy period of time.

Unluckily, there is no clear information on how the Moon was formed. Instead, there are theories and opinions: An important shortfall of the first three theories; the Co-formation Theory, the Earth Capturing the Moon Theory, and the Fission Theory, is their incapacity in explaining the high angular momenta of the Earth and the Moon. Consequently, the "Giant Impact Theory" has been the most recognized and widely accepted theory of our times. According to this theory, one of the pieces of rocks spreading out into the space after the collision of the celestial body named "Theia", which was the size of Mars, with the Earth, formed the Moon four and a half billion years ago. Modern science went back and forth from one theory to another. However, nowadays, as the origin of our Moon theory, science embraces the same process which gives the outer planets their multi-moon systems. One hindrance that is yet to be overcome lies in this question: Why did a too small Earth present itself with a single and too big Moon instead of several smaller moons?

Researches point out to the fact that the terrestrial inner planets do not have natural satellites. Everyone accepts that the two little bodies circulating around Mars are captured asteroids rather than satellites. The conditions in the Solar System are not suitable for any of the planets between the Sun and Mars to have satellites through any of the methods described. In other words, Mercury, Venus, Earth and Mars should not have had any satellites.

Nevertheless, the fact is that when we look at the sky we see a Moon that is too big for the Earth. Not only it is one eight of its mass, but also one fourth of its diameter. This ratio is exceptional in comparison to the other ratios in the Solar System.

The calculations for the oldest Lunar rocks point out to 4,25 billion years, and 4,6 billion years for the soil from the Moon. All of the approximately fifteen hundred scientists, who have examined the lunar rocks and soil brought back to Earth, agree that the age of the Moon coincides to the times when the Solar System was formed in its earliest structure. In other words, it is believed that the Moon was formed about thirty to fifty million years after the Solar System. But, then, four billion years ago something happened!

For the answers, we will go back to the ancient times. According to Enuma Elish, which originates in Sumer, but was re-written in Babylon, the present location and status of the Moon has to do with the Celestial War between Nibiru (Marduk) and Tiamat approximately 4 billion years ago, not the early formation of the Solar System 4,6 billion years ago. In the Solar System, first the Sun was formed, then Tiamat and Mercury, followed by Mars, Venus, Jupiter and Saturn. Pluton was born as a satellite of Saturn, and Uranus and Neptune were formed. The orbits were not orderly yet, and Tiamat's satellite Kingu (The Moon) was on its path to becoming a planet. Based on this, the Moon had appeared not as Earth's satellite, but rather Tiamat's, which was bigger. The Sumerians ascribed a series of satellites "totaling to eleven" to Tiamat. They placed Tiamat further than Mars, and the satellites it adopted were not formed in any different way than the ones for the other outer planets. While the formation was on, Nibiru, which belonged to another system, entered the network of Neptune in the clockwise direction, and it was pulled further into the system.

Nibiru forms seven satellites while passing by Neptune, Uranus, Saturn and Jupiter. Tiamat, which was between Mars and Jupiter, begins waiting for it together with its eleven satellites the biggest of which was Kingu (The Moon). As Nibiru approaches, Tiamat slows down, and cracks appear on the surface. The two planets coming from opposite directions were about to pass tangential, but one of the satellites of Nibiru collides with Tiamat, and divides it into two. Tiamat not only gets divided into two, but also loses all of its satellites other than Kingu. On the other hand, Nibiru gets its new route of 3600 years that would last forever. It becomes the outermost planet of the system in an elliptical orbit. Upon this collision, the seed of life flows from Nibiru onto Tiamat. During its next passage, Nibiru collides with one of the two halves of Tiamat, and casts it together with its satellite Kingu between Mars and Venus. Then, in the following passages, it shatters whatever is left down to smithereens. These smithereens have formed the asteroid belt. Simultaneously, Pluton escapes Saturn, and gets the status of a planet. The VA 243 tablet on display in the Berlin State Museum gives us the sequence of the planets. Those who wonder the details of this celestial war can read Zecheria Sitchin's book "The 12th Planet".

The Traces of the Anunnaki on the Moon

Those of you who have watched the movie "2001: A Space Odyssey" will remember; the main object in the movie was a giant monolith, that is; a block of stone. In this timeless novel written by Arthur C. Clarck, and the movie based on the book, this monolith guides humanity. How interesting it is that the monolith, which was sent by the aliens to the Earth, and which ignites the transformation to the Homo Sapiens in the movie, have been factually observed to exist among the celestial bodies.

George Leonard Ranger, author of "Somebody Else Is On The Moon", has mentioned that a similar monolith was seen near the crater, but then, had disappeared, and appeared again in other places, only to disappear once again. Researcher George Leonard has showed as evidence, a photograph he had received from a NASA staff member, whose name he keeps confidential. He has also made the giant block of stone near the same crater public on TV screens. The image is quite clear, and there are figures resembling the letters Y and Z on them. Leonard continues his explanation as such: "The Moon belongs to UFOs. They have been watching us since the Bronze Age. They observe our politics and wars. They had actually left so many traces on Earth. Ranger 7's photographs were showing these." (21)

If Leonard refers to the pyramids, the ziggurats, and the megalith structures, which are not man-made, when he says "...watching and eaving several traces on Earth..." we champion the idea that these structures belong to the Anunnaki. Although we agree with Leonard, we can say that they have been watching since much more ancient times than the Bronze Age; since the times humans have come to being.

Almost every single American astronaut have stated that he or she has felt a change of opinion virtually with a spiritual characteristic about themselves, the other people, and the existence of intelligent life beyond Earth. Gordon Cooper, who had been the pilot of Mercury 9 in 1963, and the co-pilot of Gemini S in 1965, returned with the belief that "intelligent, extraterrestrial life had visited the Earth in the past ages", and he started dealing with archaeology.

Even these two explanations alone, in fact, tell this: Whatever exists on the Moon, there are similar ones on Earth. Let's think for a moment; what would an astronaut do, if he/ she saw pyramids on the Moon, and he/ she was asked to keep it to himself/ herself as classified information? He/She would keep this secret confidential, but begin to research into the pyramids on Earth as soon as he/she arrives, and be interested in archaeology. In other words, we always end up at the point we have mentioned earlier: The future is hidden in the past.

Some astronauts not only adopted an interest in the ancient past, but also in the information passed on through legends, and the religious opinions.

Ed Mitchell, one of the Apollo 14 astronauts, said "I can say that things are happening out there." Jim Irwin (Apollo 15) had been profoundly stirred, too. In his own terms, he said he felt the "existence of God", and when he was back, he climbed up Mount Ararat in search of Noah's Ark. Allegedly, Neil Armstrong, who was the first man ever to step on the Moon, went on a voyage of discovery in the same manner, in search of Noah's Ark. One should ask whether these scientists came to Mount Ararat thinking that the Myth of Noah and The Flood was actual. If that is the case, can we conclude that the actuality of the information provided to us through ancient history is gradually coming into the light?

In the last century, several researchers, who think that Noah's Ark landed on Mount Ararat, went through great troubles to carry out meticulous in situ research in our country (Turkey). Although as a nation we consider the Flood as a religious myth or a legend, the World approaches this matter in a different way. In 2010, a Chinese group announced that they had found Noah's Ark, and they presented the evidence to the scientific world during a press conference. (22)

Although we do not know the exact number of researchers who came to Mount Ararat, we learn from Hasan Ozer, who lives in Dogubeyazit, and who volunteers in familiarizing us with the traces of Noah's Ark, that the numbers exceeds hundreds. (23) All of the visitors are very serious researchers. Hasan Ozer had guided several people from all walks of life including scientists from various platforms, ancient history researchers, archaeologists, and UFO researchers in the region. He says lots of evidence was discovered, but, unluckily, some of it was stolen from where it used to be. He claims that he personally prevented several of them from being taken away. He had applied for an authorization from the Ministry of Culture, which would allow him to collect this evidence, only to eventually return them to the Ministry, which he did.

Is Noah's Ark really on Mount Ararat? If we consider the conclusions derived from the Holy Books, the notion that it is on Mount Ararat gains weight. Those who do not agree, point to Mount Judi. For the time being, it is not possible to conclude exactly and clearly on this matter. However, ancient inscriptions talk about this subject clearly. Ziusudra in Sumer and Utnapishtim in Akkad tells the story in details including Noah's construction of the Ark and the Flood itself. We think that this ship, which we come across with in many ancient texts including the Epic of Gilgamesh, has landed on Mount Ararat following the Great Flood that happened circa 11 thousand BCE. Right, but, what is the opinion of the scientific world on this matter?

Did the Deluge really happen?

In almost every culture, there are various narratives about the Deluge, which is said to have taken place 120 shars after the arrival date of the Anunnaki on Earth. So, what could this Deluge, which had a place in all the ancient nations' memories, and which finds a place in all the existing mythologies, be?

Mainstream scientists think that in every culture's past there are flood-like inundations of water, and that this is interpreted as the Deluge (the Great Flood) in that culture. According to them, a flood so huge to cover the mountains at one time, and a consequent cataclysm have never happened. However, if we assume that the Great Flood was real, the most rational date for this to happen would be the end of the Last Ice Age. Modern science dictates that the last Ice Age came to a sudden end about twelve to thirteen thousand years before now in the southern hemisphere, and about one or two thousand years after this in the northern hemisphere.

The scientists, who have drilled holes in the Arctic and Antarctic ice sheets, measure the oxygen left between different strata, and thus, they can determine the climates prevailing thousands of years ago. Specimens gathered from the bottoms of the seas such as the Gulf of Mexico are used in calculating the amounts of decreases and increases in the number of sea organisms. Through this process scientists can similarly estimate the temperatures in the past centuries. Based on such data, scientists also estimate that the last Ice Age had started seventy-five thousand years before now, and that a small increase in temperature appeared at a period about forty thousand years before now. Approximately thirty-eight thousand years ago, a harsher, colder, and drier period was endured. Then, at about thirteen thousand years from now, the Ice Age suddenly came to an end, and the moderate climate that we are enjoying now moved in. Among all the theories developed by the scientists, one stands out for us. This theory was suggested by Dr. John T. Hollin from the Maine University. Hollin has reached the conclusion that the Antarctic ice sheet slid into the sea creating a sudden large-scale tidal wave as it kept on breaking down periodically.

This hypothesis suggests this: As the ice sheet kept on getting thicker, not only it confined more and more of Earth's heat underneath as time went on, but also, it created a diluted and slippery stratum on its base. This diluted icy stratum, which served as a lubricant, between the thick, upper ice sheet and the solid soil at the bottom layer, sooner or later, caused the ice sheet to slip towards the ocean surrounding it.

Hollin had calculated that the level of all the seas of the Earth would rise by approximately 20 meters, in case just half of the existing Antarctic ice sheet slides into the South Seas, causing a tidal wave to cover all the cities on the shores and all the lower areas

In 1964, A. T. Wilson from the Victoria University in New Zealand, suggested that the ice ages came to a sudden end as a result of such slides not only in the Antarctic, but also in the Arctic. In the light of all this information, we think that the Deluge happened as a result of billions of tons of ice sliding into the waters of the Antarctic, which in return suddenly brought an end to the last Ice Age. This sudden event caused an enormous tidal wave. Starting in the waters of the Antarctic, it expanded in the direction of north to the Atlantic, Pacific and Indian oceans. The sudden change in temperatures must have created torrential storms accompanied by flooding rains. The storms, the clouds and the darkening skies had acted much faster than the waters, and they had given an alarm about the approaching flood. The ancient texts describe exactly such a phenomenon.

The Sumerian texts connect the reason triggering the Deluge to Nibiru's passage on that date. This outermost planet which passes near the Earth every 3600 years, had a much larger effect during its passage coinciding to the times circa eleven thousand BCE. The reason for this is explained through all the inner planets taking stand on the other side of the Sun, thus exposing the Earth directly to Nibiru, and making it feel all of Nibiru's gravity all by itself.

The researchers of our times do not give a warm welcome to the idea of Nibiru yet. Instead, they consider the triggering cause to be a comet either passing too close by the Earth or colliding with it. (24)

If we go back to our main subject, we hear Edward G. Gibson, who was a scientist in Skylab 3 (1974), saying "To be in the orbit of the Earth for days at a time, drives you to speculate a little more on the life that exists in the other parts of the universe." Those who have been particularly affected are the astronauts who had visited the Moon.

On the 20th anniversary of the first landing on the Moon, Al Worden, the Apollo 15 astronaut, appeared on a TV program in which he likened the moon module used in landing on the Moon, and in taking-off vertically from the Moon, to the spacecraft described in Prophet Ezekiel's vision. Also, during mission Apollo 11, that is; during the first successful landing on the Moon, an interesting event happened. When Armstrong and Aldrin took the first steps on the surface, Michael Collins was in the command module named Columbia, high up in the orbit, and he saw a few flying objects. Collins described and reported these objects to Earth using the term "Bogey". During this period, the conversations down below were like this:

Armstrong suddenly screamed; "What was it? What the hell was it? That is all I want to know!"

Mission Control: "What is there?

Malfunction... Garble... Mission Control calling Apollo 11...

Apollo 11: "These babies are huge, sir!... Enormous! ... Oh, God! You wouldn't believe it! I'm telling you there are other spacecrafts out there... They are here, they are parked on the side of the crater. They are watching us. They were here before us..."

In the hours following, Mission Control tells the astronauts to stop the conversation and record the objects in film.

Buzz Aldrin, the other astronaut of Apollo 11, have stated his belief in extraterrestrial life like this: "One of these days, through telescopes like Hubble, rotating in the orbit, or through other technical discoveries, we may find out that we are really not alone in this wonderful universe."

Today, the number of people, who think that there are some living on the Moon, is quite high. Allegedly, there are bases on the other side of the Moon which cannot be seen from the Earth. Major Donald Keyhoe, a UFO researcher has come to the conclusion that there are extraterrestrial astronauts on the Moon. In his book, "Flying Saucers Conspiracy", Keyhoe explains his belief so:

"All clues point out to not only the existence of a Moon Base, but also the operations of a smart race, which have already begun. If this is true, who could these entities be? Do they come from other planets, or do they originate on the Moon? We should accept that the northern territories have variety. The entities living on the surface might have settled much before the atmosphere of the Moon was thinned. In this case, they might have adopted to the conditions by creating an artificial "underground atmosphere"... But, it is more probable that the Moon Race —if there is such a race— lived underground at all times in order to escape the continuous meteor bombardments. (25)

Another UFO writer, Harold Wilkins says in his book, "Flying Saucers on the Moon": "I had outlined the possibility that the Moon could have been a stop-over place for the spacecrafts, which we call as flying objects, and that it may still be so." He adds: "Based on the suggestion of the theory that our satellite, the Moon has been used, and is still being used as an advanced observation base in reference to our World by mysterious cosmic visitors, who are responsible from the flying objects phenomena, I need to emphasize the weirdness of these light centers being on the North-west or one quarter part of the Moon." (26)

Considering all of the above, we can conclude that there is actually a base, not life on the Moon. Although there are many purposes for an underground base, its fundamental job is to observe the Earth, and to serve as a way station for the spacecrafts. It is obvious that those who use this base on the Moon are the ones who have built the ancient structures on Earth. Using inductive reasoning, the information leads us to the Anunnaki, and points out to the Moon as one of the stations between Nibiru, and Earth.

The stepping stone of Humanity: Mars

Before furthering on the subject of Mars, let us take you back to 1926, and let's see what was written in the Resimli Ay Mecmuasi (Illustrated Moon Magazine), which was being published in Ottoman language at a period when modern Turkey was already established. The article from the magazine with the title "Rehearsals for War with Mars are being held." explains this:

"Last month, a rumor traveled around London that news from Mars have arrived. Strange signs were recorded in one of the wireless telegram stations on a peaceful and quiet night. The man who recorded these signs was one of the scientists of U.K. in the field of astronomy. This scientist, who viewed these signs as news sent from Mars, wrote a telegram message himself in order to send to Mars. He took it to the wireless telegram office. He asked them to transmit it to Mars. Thousands of curious people from the public gathered in front of the office waiting for the answer, but, nothing came." (27)

We can say that we collected much more information on Mars in the period —roughly around a century— following the news on "the answer expected from the Martians", which we read with a grin on our face. The most important information on Mars was collected through the Martian Meteorite on Antarctica. Sixteen million years ago, a meteorite or an asteroid had a collision with Mars causing a piece of rock from Mars to be blasted, and it landed on Antarctica as a meteorite thirteen thousand years ago. After examining ten thousand pieces of rocks collected since that day, on August 7, 1996, NASA gave this explanation, which would set the world on fire, during a press conference: "Strong evidence have been reached pointing out to life on Mars at one time."

Due to its mass at one to ten ratio in comparison to Earth, Mars' gravity is very low. This implies that Mars is just recently going through the times Earth once went through. All this makes Mars suitable for a perfect way station. In the meantime, we can't help wondering whether the fact that the meteorite in Antarctica landed on 11 thousand BCE has to do with the Great Flood.

The Fenomen Dergisi (The Phenomenon Magazine), which started its publishing life in Turkey, in 1996, with Ata Nirun as the director has touched on many of our lives. In the fourth volume of this magazine, very special information on Cydonia, the Martian City, and the "Face on Mars" was shared. Based on the claims of an Italian writer, whose name was mentioned in "The

Mars Flight Atlas" published by Harper & Row, the American spacecraft Viking 1 have discovered an ancient city on Mars, but the information has been censored, and the subject is kept confidential. In one of the photographs Viking 1 took, and transmitted to Earth on June 19, 1976 after entering the orbit of Mars, there is the image of a human face with a slightly open mouth, and an eagle-head helmet on, whose eyes are directly aiming at the spacecraft. Dr. Tobias Owen, who commanded the flight crew of Viking 1 from Mission Control on Earth, is the one who noticed this humanoid face. Dr. Owen has published a book called "Face on Mars" about his finding, and has hit record sales. NASA says that this phenomenon, which is known as "The Face on Mars" since that day, is nothing but a play of light and shadows. NASA also showed the photographs taken later by the spacecraft in which the face is not visible. However, despite all the efforts, NASA's explanations have not been convincing, because the later photographs taken by the craft are questionable.

Had the Face on Mars been a singled out phenomenon, perhaps, it would have been long forgotten. However, the photographs reveal many more mind-boggling shapes, and an ancient city of perfect mathematical ratios. The area in mention have been named as Cydonia, and this name became famous as the Ancient Martian City: Cydonia. Vincent Di Pietro, and Gregory Molenaar, who are employed on a contract at the Goddad Space Flight Center, which is a NASA institution on the Eastern shores of Maryland, brought out the photograph that was sinking into oblivion, and examined it by using modern image composition techniques. In their book named the "Unusual Martian Surface Features", one can see the eyeballs in the eye sockets, and the teeth in the mouth on the Face on Mars. In the same period, a book named "Us, Mars", which was published in German, makes a connection between the image on Mars and the Sphinx in Egypt. While the discussions continued on one side, Dipietro and Molenaar described a giant pyramid expanding on an area of at least two to three square kilometers at a distance of approximately twenty kilometers from the Face on Mars. This pyramid is named after the two scientists using their initials: D&M. Richard Hoagland, the ex NASA consultant, and popular science writer, defines this pyramid as an object similar to the Rosetta Stone, which is the key to the hieroglyphs. According to him, some kind of a mathematical code is hidden in the calculation system of their internal angles, and the answer to this puzzle lies in Cydonia. In his book named "The Monuments of Mars", Hoagland mentions that a second face appears after high computer scanning on two photographs. Besides, there are fifty meter mounds, and a shape resembling to the Star of David. The photo of the castle on the cover of the book "The Martian Enigmas: A Closer Look" written by Mark Carlotto, is really very impressive.

A report covering the Cydonia research have been published in 1993. In this report, Dr. Stanley McDaniel, Professor of Philosophy Emeritus and former Chairman of the Department of Philosophy at Sonoma State University in California says this: "There is a crucial mistake; the U.S. Congress, and the scientific institutions are responsible for this. They should have been kept liable for researching the structures on the surface of Mars, which are possibly products of intelligence, because right now, we are in the Mars Discovery Program, and we are working with huge costs. This matter must be clarified."

Prof. McDaniel has presented a report to the Stanford University Scientific Discovery Alliance, working together with Dr. Horace Crater from the University of Tennessee Space Institute. The report includes the statistical results of studies that have been carried out for a long time, and computer research. The most important conclusion of the report is that the so called "City Center" in Cydonia cannot be natural, or seen coincidently the way it is, as proven by the studies on the angles, and mathematical image examinations. (28)

Is Mars going through Ice Age?

Nowadays, it is believed that the ice ages of Earth are caused by the three main oscillations of Earth's orbit around the Sun. The first one is the shape of the orbit itself: It has been concluded that the orbit has changed from a circular form to a more elliptic one in a period about hundreds of thousand years. This change causes the Earth to be sometimes closer, and at other times further than the Sun. Secondly, the Earth's axis is tilted 23.7 degrees from the plane of its orbit around the Sun, but this tilt is not constant. During a cycle that averages about 41 thousand years, the tilt of the axis varies between 21 and 24 degrees. More tilt means more severe winters and summers, and a change in air and water currents. The third phenomenon is the precession of the equinox; Earth's axis draws an imaginary circle as the Earth rotates around itself while tilting to left and right, and it takes 25920 years to complete this cycle.

Planet Mars goes through these three cycles, too. However, its greater orbit around the Sun, and the greater tilt change creates even more severe climate changes. As mentioned before, it is believed that this cycle takes about 50 thousand years on Mars. When the next moderate period of Mars arrives, perhaps, the planet will literally be full of flowing waters, the seasons will not be so severe, and its atmosphere will not be as challenging as it is now for human beings. We believe that the last "moderate" period on Mars must have been quite recent, because had it not been the case, the dust storms of Mars would have completely covered the evidence of the rivers once flowing on the surface, the ocean shores and the lake basins, and there would not be even the amount of vapor that exists in its atmosphere today.

Harold Masursky from the U.S. Geological Survey said, "Speaking from a geological point, there were flowing waters on the Red Planet." We think that this change happened simultaneously with the Earth, that is; around 11 thousand BCE. While triggering the Great Flood on Earth, it took away the atmosphere of Mars, and caused the Martian meteorite to land on Antarctica. Although it is not scientifically referenced, this phenomenon has to do with Nibiru's passage on that date, which is mentioned in Sumer. Nibiru, which completes its passage between Mars and Jupiter every 3600 years by approaching the Sun because of its elliptic orbit, caused various catastrophes in Mars, the Moon and the Earth. Specifically, during its passage on 11 thousand BCE, even greater catastrophes were experienced due to several variables occurring at the same time.

What makes the mystery even deeper is the astronauts' discovery of the surface shapes upon seeing the photographs of the surface of Mars, which made them give the name "The Inca City". These shapes located on the southern part of the planet show a series of upright walls formed of square-like or rectangular pieces. John McCauley, who is an ex NASA geologist, has commented that the "bumps are continuous, not showing any sign of break points, and standing up distinctively like the walls of an ancient ruin in between the plains and small hills around" This amazing wall or structured series of stone blocks has an awe-striking resemblance to great and mysterious structures such as those in Baalbek, Lebanon and Sacsayhuaman in Cusco, Peru.

Considering the resemblance of the structures of Earth and Mars, one of the rational alternatives is that some entities out there, who were capable of space travel, had visited both Earth and Mars about half a million years ago, and had left behind monuments. The only entities whose existence can be evidenced in the Sumerian texts, the holy texts and all the ancient mythologies are the Anunnaki coming from Nibiru. After the atmosphere of Mars vamoosed around 11 thousand BCE, these structures were deserted just like our Great Pyramids. The Anunnaki, whom we think are keeping on with their activities underground in Mars, will perhaps come out to the surface again during the next Martian period of moderate climate. A short while after arriving on Earth, the Anunnaki have also stepped on Mars. As only a very limited amount of equipment or material could be carried from Nibiru, the structures on this planet were built out of stones and rocks, which was the hardest of its own equities, just like on Earth. The planet, which has served as a way station between Nibiru and Earth for many years without any problem, has a structure that allows life, especially around its equator. The water of the planet is drinkable, but there are no live organisms.

If we consider Zecheria Sitchin's views; there was an environment fit for life on the surface of Mars before the catastrophe. The structure carved on the rock that we consider as The Face on Mars in the city of Cydonia, which is about one and a half kilometers long, and eight hundred meters high from the plain, is a burial place for Alullim, who is the first person in the Sumerian Kings List.

The orientation of the Face on Mars and the Great Pyramid in Cydonia has been constructed in alignment with the equinox of the Sun on Mars. This was designed so that Alalu could face the Sun. Just like the Anunnaki cities on Earth that were built with consideration to Mount Ararat, the cities and landing areas here were built in reference to Mount Olympus Mons, which is twenty-four thousand meters high. Cydonia's pyramids, the landing area in the middle, its sphinx and the agricultural lands in this area can clearly be seen. Cydonia, which is located about forty-five degrees above the equator, served as the center.

The Martian base have served the Anunnaki for thousands of years with Elysium, which is one hundred and fifty degrees to the east, and adjacent to the equator; Utopia, which is thirty degrees north of Elysium; and the city of Inca, which is far from the center. Mission Mars has been planned for limited areas, with limited means, only to serve the purpose. Elysium attracts attention as a city expanding in a rhombus shape, which includes a landing area, numerous small structures and pyramids in the middle of two pairs of tetrahedron pyramids two of which are very big and the other two smaller. It is estimated that Elysium was used as the agricultural and irrigation region. The Elysium pyramids and agricultural lands are observed next to each other. Along the shores of the lake to the south of the Face on Mars five gold depots and a dock can be seen. In Utopia, there is a runway in front of a pentagonal flight center. Utopia served as the flight control center for the space port in Cydonia. Utopia's runway, which was lined on the hard ground, seems to have posed no difficulty for the celestial discs during landing and take-off. The underground silo, the flight signs and the other structures can be seen even today.

The Monolith of Phobos and the Human Beings meeting with the Ancestors: The Accident of Phobos

Phobos is one of the two tiny asteroids rotating around Mars. The other one is Deimos. How they got into their orbits is still a mystery. Although several theories have been projected, the latest theory that has been accepted explains that Mars used to have many small satellites long time ago. However, over time, all of them landed on Mars, and only Phobos and Deimos remained. According to this theory, in about thirty to fifty million years, these two satellites will also land on Mars.

Both asteroids are very small and irregular in shape. What makes Phobos precious for us is the evidence which makes us believe that there is still life underground. NASA astronaut Buzz Aldrin, the second man to walk on the Moon, has talked about a monolith; a strange and single standing big piece of rock sitting on the surface of Phobos, on a television program he appeared on in 2009. The monolith has been described as a ninety-meter big rock or as big as a building. What makes the Monolith of Phobos more impressive is its location in a solitary area. This monolith, which has been the subject of discussions on Reddit with the conspiracy theories, has landed its name to an album released in 2016 by Les Claypool and Sean Lennon Ono. The

Monolith of Phobos has not been the subject of scientific interest as of yet. (29) However, we can see, in retrospect, that it is very meaningful; perhaps it is a message for us.

After 1985, very big changes have been happening in the World. It seems as if a magical hand from above has intervened, and changed something in a way we cannot comprehend. These changes have affected the entire World and have peaked in 1988. The trigger of the change is USA, but its center is USSR. Not only in USA, but in the entire World, a new era started with the announcement of ceasefire between Iran and Iraq; Benazir Butto's election as the first female prime minister in Pakistan; George W. Bush's inauguration as president in USA in 1989 following Ronald Reagan. In this new era, following USSR's complete withdrawal from Afghanistan, the Eastern Block countries were on the verge of boiling and the Communist Party was defeated at the first free elections in the USSR.

In August, two million people living in Letonia, Estonia and Lithuania formed a human chain about six hundered and fifty kilometers long by holding hands, demanding freedom and independence from USSR. On October 3, the government of Eastern Germany closed its borders with Czechoslovakia in order to stop the westbound migration. While Hungary and Czechoslovakia was successfully working on separation from the Eastern Block, the Berlin Wall was being torn down. By the end of the year, consequent to a meeting in Malta, US President George H. W. Bush and Soviet leader Mikhail Gorbachev announced that the Cold War is over. In the periods following, events like the wars in the Balkans and Middle East, dissolution of USSR and Eastern Block countries' proclamation of independence took place.

It is as if an era came to an end in 1988, and a new one started in 1989. Historians, sociologists, and political scientists have assessed this five year period from every angle, and each has given an explanation from his own point of view. However, none has been satisfactory for the World. We think that starting with 1989 there has been an intervention on Earth from outside. In his book "The Cosmic Code" Sitchin proposes that the Phobos Accident, which happened on March 27, 1989, was actually an intervention by the Anunnaki. One piece of information that validates this proposition, which we agree with, is the information provided to the western media two years after the incident by the Russian astronaut and pilot Colonel Dr. Marina Popovich. During a UFO conference organized in 1991, Popovich shows a very important photograph besides some very interesting information "leaked" from the Soviet Union. (Photo 1)

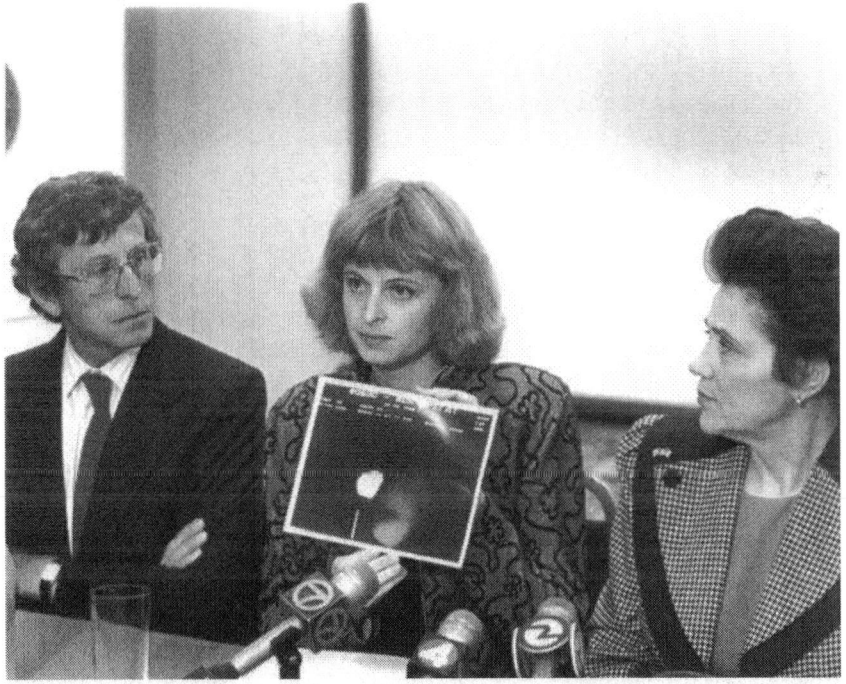

Photo 1: Dr. Marina Popovich

In this photograph taken two days before contact was cut off with Phobos 2, that is; on March 25, 1989, the big "main ship" in the shape of a cigar, which was 20 kilometers long, with a diameter of 1,5 kilometers, was parked or hanging in the air next to Phobos.

This ship had mysteriously disappeared after the Soviet unmanned probe Phobos 2 had transmitted this photograph. Rumor had it that this main ship had probably struck Phobos 2 with a shock wave or an energy blow. (30)

We are suggesting that the cigar ship is an Anunnaki freight ship, and Phobos is their base. Following the incident of Phobos 2, this matter of threat suddenly lost its popularity, and Phobos, just like the Moon, was pushed into the background.

The year of 1989 has made it clear that, by all means, it is the beginning of a new era. In the period following it, it is as if a hidden hand has started reorganizing the political system of the World. This hidden hand, which initially used to say that it is of no threat, later on proved that everything actually develops under its control.

The proposition that Phobos is an asteroid that can be used as a "spacecraft hanging in the orbit" is being discussed in the scientific environments. It is easy to go back and forth to Phobos, whose distance to Mars is one third of the Moon's distance to Earth. The very weak gravity of Phobos makes it possible for the spacecrafts to land on Phobos by putting on the brakes briefly, and to take-off by using very little force. The ice underneath the surface of Phobos can be converted to drinkable water, and then separated into oxygen and hydrogen. The hydrogen can be combined with the carbon atom to create hydrocarbons. Research can be carried out for the uses of oxygen and hydrogen, using them as fuel to be the first. It has been asserted that many strategic minerals including chromium, germanium and gallium exist on Phobos. It is possible to turn Phobos into a self-sustaining space base through some good planning.

Consequent to all this, we believe that there is an Anunnaki space base on Phobos. Phobos, which completes three and a half rotations in one day around the planet, and which can be easily seen from all but one of the Martian cities, is always in front of the eyes. We can envision that Phobos might have been turned into a much lighter asteroid by emptying it from the inside and protected from the cold and radiation of the space with its crust empowered further with an added shield of protection. It is possible that the craters serve as gates while coming out from the inner cities. We can easily see even today that the channels created around the asteroid are connected to these craters. Furthermore, we think that Enoch visited Phobos, too.

Sources:

1. Fenomen Dergisi, 1996 4. Sayı, 60. Sayfa
2. https://tr.wikipedia.org/wiki/Mars_One
3. Vance, Ashlee, Elon Musk, 4. Baskı, Buzdağı, 2017, İstanbul, S.385
4. http://www.ntv.com.tr/teknoloji/marsa-yolculuk-icin-80-bin-basvuru,EhQuKgYwNUGFdAiBJS5oLw (28.07.2017)
5. http://www.milliyet.com.tr/-mars-kolonisi-nde-turkler-gundem-2015489/ (28.07.2017)
6. http://etcsl.orinst.ox.ac.uk/cgi-bin/etcsl.cgi?text=t.1.1.4&charenc=j#
7. http://etcsl.orinst.ox.ac.uk/cgi-bin/etcsl.cgi?text=t.1.5.1&charenc=j#
8. http://etcsl.orinst.ox.ac.uk/cgi-bin/etcsl.cgi?text=t.1.6.1&charenc=j#
9. http://www.haberturk.com/saglik/haber/1453245-kafa-nakli-ne-zaman-yapilacak (20.07.2017)
10. https://tr.wikipedia.org/wiki/Bengi_su
11. http://www.hurriyet.com.tr/marsta-hayat-daha-uzun-39015830
12. http://www.ntv.com.tr/turkiye/insanligin-omru-uzay-yolculugu-ile-uzayacak,8DBrEPppqU6hpiBVfbMzEA
13. http://www.ntv.com.tr/galeri/teknoloji/marsa-yerlesen-insanlar-evrim-gecirecek,dVWJ6JkbQU2CFYuw7eQDyQ/ux-hqg7s7U-u1sp-rpl04Q
14. https://onedio.com/haber/jupiter-in-uydusu-europa-da-yasam-ihtimali-guclendi-713023
15. http://www.milliyet.com.tr/titan-da-yasam-belirtileri-magazin-1246792/
16. http://www.ntv.com.tr/galeri/teknoloji/nasadan-beklenen-aciklama-saturnun-uydusunda-yasami-destekleyebilecek-deliller,wW6SDvpeTkuckgp_qzLafg/xdNhpAvvB0SSR23Q-zgZRA

17. http://www.hurriyet.com.tr/ay-neden-bombalandi-12752358
18. https://tr.wikipedia.org/wiki/Ay
19. https://tr.wikipedia.org/wiki/Ay
20. http://web.archive.org/mission_pages/LRO/news/apollo-sites.html
21. http://web.archive.org/web/20140318041834/http://lroc.sese.asu.edu:80/news/index.php?/categories/2-Featured-Image
22. http://web.archive.org/web/20150725152127/http://www.bbc.co.uk/news/science-environment-19050795
23. https://www.uzayyolu.net/ay-da-cok-buyuk-bir-gariplik-var/
24. http://www.hurriyet.com.tr/nuh-un-gemisi-agri-daginda-bulundu-14542202
25. https://www.sondakika.com/haber/haber-nuh-un-gemisi-nin-izlerini-30-yildir-gonullu-5195693/
26. http://www.barry.warmkessel.com/3related.html
27. http://www.ufologism.net/ay_bolum_2_haberi_65.html
28. http://www.ufologism.net/ay_bolum_2_haberi_65.html
29. https://www.haberler.com/1926-yilindaki-osmanli-mecmuasinda-mars-ta-insan-8157032-haberi/
30. http://posthumanblues.com/mactonniescom/imperative42.html
31. http://www.bbc.com/earth/story/20160923-there-is-a-huge-monolith-on-phobos-one-of-marss-moons
32. https://www.bibliotecapleyades.net/marte/marte_phobos05.htm

SECTION 5

ENKI - ADMIRAL BYRD MEETING

Just as the long night of the Arctic ends, the brilliant sunshine of Truth shall come again... and those who are of darkness shall fall in its Light... For I have seen that land beyond the pole, that center of the great unknown.

Admiral Richard E. Byrd

US Navy December 24, 1956

Was it Enki who met with Admiral Byrd?

Having traced the footsteps of the Anunnaki in the skies, we wonder if they might have bases on Earth, too. Could there be Annunaki bases under mystic places such as the Poles, the Sinai Peninsula and the Gobi Desert? Are legendary cities like Agarta and Shamballa surface or underground bases on Earth? We believe that the interesting and mysterious incident Admiral Byrd experienced will shed light for us.

Richard Evelyn Byrd is a successful US naval officer, who joined the air service of the US Navy at age twenty-four, and climbed up to the rank of admiral. During his life as an officer, Admiral Byrd flew over unknown places such as the North Pole, Greenland, and Antarctica; and he carried out confidential scientific studies in these regions in the name US Navy. Despite being such an acclaimed officer, the World knows him through the secret information he shared at a time close to his death. This is how he explains why he kept the information to himself for many years, and why he gave it to the world three months prior to his death risking his reputation as a hero: (1)

"I have kept this secret for many years. This was against all my moral values and rights. Now I feel that the eternal night is here, and this secret should not die with me. But, reality will eventually prevail. This is the only hope for humanity. I fulfilled my duty for the industry that is the heart of the military monsterity. Now the long night is starting, but this will not be an end."

Prof. Dr. William Bernard, who published Admiral Byrd's memoirs, explains whether the experience was a hallucination or not in this way:

"These memoirs were written in February and March of 1947. The conditions which the Pole Explorer Admiral Byrd was in were endurable and safe. Different people have provided assurance that it was not a matter of hallucination. What is written consist of verbatim words of the Admiral. They were written on a long night at the North Pole, and it tells the reality experienced by a serious explorer and a scientist under bright day light.

Admiral Byrd is very important for us, for we think that he met with Enki in February of 1947. Although it seems like Enki was warning the US government about two atomic bombs by using the Admiral as an emissary, on a deeper scrutiny it becomes clear that the messages were sent to all humanity. Furthermore, we think that Enki added in the coded messages in between the lines, which can only be seen by the awakening people. Well, in this section, we will be tracking the traces of Enki's base in the North, and the messages he gave to the awakening people by examining the intriguing experience of the Admiral. Now, let's see what the secrets of the confidential diary of the Admiral were:

Admiral Richard B. Byrd's Diary, February - March 1947

"The Inner Earth – My Secret Diary
The exploration flight over the North Pole

I must write this diary in secrecy and obscurity. It concerns my Arctic flight of the nineteenth day of February in the year of Nineteen and Forty Seven.

There comes a time when the rationality of men must fade into insignificance and one must accept the inevitability of the Truth!

I am not at liberty to disclose the following documentation at this writing... perhaps it shall never see the light of public scrutiny, but I must do my duty and record here for all to read one day.

In a world of greed and exploitation of certain of mankind can no longer suppress that which is truth.

Flight Log – Base Camp Arctic – 2/19/1947

0600 Hours- All preparations are complete for our flight northward and we are airborne with full fuel tanks at 0610 Hours.

06:20 Hours- Fuel mixture on starboard engine seems too rich, adjustment made and Pratt Whittneys are running smoothly.

07:30 Hours- Radio Check with base camp. All is well and radio reception is normal.

07:40 Hours- Note slight oil leak in starboard engine, oil pressure indicator seems normal, however.

08:00 Hours- Slight turbulence noted from easterly direction at altitude of 2321 feet, correction to 1700 feet, no further turbulence, but tail wind increases, slight adjustment in throttle controls, aircraft performing very well now.

08:15 Hours- Radio Check with base camp, situation normal.

08:30 Hours- Turbulence encountered again, increase altitude to 2900 feet, smooth flight conditions again.

09:10 Hours- Vast Ice and snow below, note coloration of yellowish nature, and disperse in a linear pattern. Altering course foe a better examination of this color pattern below, note reddish or purple color also. Circle this area two full turns and return to assigned compass heading. Position check made again to base camp, and relay information concerning colorations in the Ice and snow below.

09:10 Hours- Both Magnetic and Gyro compasses beginning to gyrate and wobble, we are unable to hold our heading by instrumentation. Take bearing with Sun compass, yet all seems well. The controls are seemingly slow to respond and have sluggish quality, but there is no indication of Icing!

09:15 Hours- In the distance is what appears to be mountains.

09:49 Hours- 29 minutes elapsed flight time from the first sighting of the mountains, it is no illusion. They are mountains and consisting of a small range that I have never seen before!

09:55 Hours- Altitude change to 2950 feet, encountering strong turbulence again.

10:00 Hours- We are crossing over the small mountain range and still proceeding northward as best as can be ascertained. Beyond the mountain range is what appears to be a valley with a small river or stream running through the center portion. There should be no green valley below! Something is definitely wrong and abnormal here! We should be over Ice and Snow! To the portside are great forests growing on the mountain slopes. Our navigation Instruments are still spinning, the gyroscope is oscillating back and forth!"

The Admiral says that a small river is running right in the middle of the North Pole, and an all green valley lies down below him, and he questions its rationality.

"10:05 Hours- I alter altitude to 1400 feet and execute a sharp left turn to better examine the valley below. It is green with either moss or a type of tight knit grass. The Light here seems different. I cannot see the Sun anymore."

We understand from the words of the Admiral that this place, which is isolated from our current world, has its own unique light source: "The Light here seems different. I cannot see the Sun anymore." Do you remember the French animation movie broadcasted on television as a fifty-two episode series in the years between 1985 and 1987? This animation movie, "Les Mondes Engloutis" was broadcasted on TRT (Turkish Radio Television) under the title "The Lost Worlds", and its plot was quite riveting:

Thousands of years ago, there was a civilization with a highly advanced technology on Earth. Following the Great Disaster, this civilization had collapsed, and the city of Arcadia was buried in the center of the World. For years, the inhabitants of Arcadia believed that they were the only ones saved from the Great Disaster. Therefore, they re-established their civilization in the center of the World. There, they created an artificial sun named Tehra. (2)

The greatest resemblances between this animation movie and what the Admiral tells are the artificial source of light, and the underground civilization, which we will see in a bit. If we agree that there is an Anunnaki base in the North, then, we need to ask: "Who is the owner of this base, and what is his mission?" For the answer, let's continue with what the Admiral says:

"We make another left turn and we spot what seems to be a large animal of some kind below us. It appears to be an elephant! NO!!! It looks more like a mammoth! This is incredible! Yet, there it is! Decrease altitude to 1000 feet and take binoculars to better examine the animal. It is confirmed - it is definitely a mammoth-like animal! Report this to base camp."

We see that the mammoths, an extinct species, are taken under protection in this place, and we catch the first hint. Through the Flood Tablets, we know how much importance Enki gives to animal types. If the mammoths are still alive, they must have been taken under protection in this area. In the meantime, let us keep in mind the mammoth, which is said to be seen in Siberia in 2012. (3)

"10:30 Hours- Encountering more rolling green hills now. The external temperature indicator reads 74 degrees Fahrenheit! Continuing on our heading now. Navigation instruments seem normal now. I am puzzled over their actions. Attempt to contact base camp. Radio is not functioning!

11:30 Hours- Countryside below is more level and normal (if I may use that word). Ahead we spot what seems to be a city!!!! This is impossible! Aircraft seems light and oddly buoyant. The controls refuse to respond!! My GOD!!! Off our port and starboard wings are strange types of aircrafts. They are closing rapidly alongside! They are disc-shaped and have a radiant quality to them. They are close enough now to see the markings on them. It is a type of Swastika!!! This is fantastic. Where are we! What has happened? I tug at the controls again. They will not respond!!!! We are caught in an invisible vice grip of some type!

Then, the controls of the plane are locked by a weird flying disc. It is intriguing that the Swastika sign, which is one of the "Mu" symbols, appears on this aircraft described as "disc-shaped and have a radiant quality" At this point, we remember the Book of Ezekiel in The Torah. Ezekiel's meeting with Yahweh in Section 1:4/5 was told so: *"I looked, and I saw a windstorm coming out of the north—an immense cloud with flashing lightning and surrounded by brilliant light. The center of the fire looked like glowing metal, and in the fire was what looked like four living creatures, in appearance their form was human...*

11:35 Hours- Our radio crackles and a voice comes through in English with what perhaps is a slight Nordic or Germanic accent! The message is:

> *'Welcome, Admiral, to our domain. We shall land you in exactly seven minutes! Relax, Admiral, you are in good hands.'*

I note the engines of our plane have stopped running! The aircraft is under some strange control and is now turning itself. The controls are useless."

Swedish or Germanic accent reminds us of the US-Anunnaki encounter in 1954. There, the Anunnaki were likened to the Nordics, that is; the northern European societies.

"11:40 Hours- Another radio message received. We begin the landing process now, and in moments the plane shudders slightly, and begins a descent as though caught in some great, unseen elevator! The downward motion is negligible, and we touch down with only a slight jolt!

11:45 Hours- I am making a hasty last entry in the flight log. Several men are approaching on foot toward our aircraft. They are tall with blond hair."

If you will recall, the extraterrestrials, whom US President Dwight D. Eisenhower had an encounter with on February 20-21, 1954, and had named as Nordic, were also described as Caucasian with blue eyes and blonde or white hair. Almost every tablet mentions that Anunnaki are taller than human beings. Additionally, in the Enki and Ninmah Creation Myth (4), there is mention of the first human being as "Adamu", who is black, and that he is different than the Anunnaki. We think that we have a hold of our second hint from "the Anunnaki are Caucasian and tall".

"In the distance is a large shimmering city pulsating with rainbow hues of color. I do not know what is going to happen now, but I see no signs of weapons on those approaching. I hear now a voice ordering me by name to open the cargo door. I comply.(END LOG.)"

From this log we understand that the admiral had arrived there with special permission. The Admiral says that he could not record what came next in the log, but he was writing out of his memory.

I am entering the crystal city...

"From this point I write all the following events here from memory. It defies the imagination and would seem all but madness if it had not happened. The radioman and I are taken from the aircraft and we are received in a most cordial manner. We were then boarded on a small platform-like conveyance with no wheels!"

The Admiral's entry into this crystal city reminds us the special place Gılgamesh had went to, to meet Utnapishtim (Noah). Gılgamesh had described what he had seen as " 'a chamber of gods' with a 'garden' in it, which is entirely made of precious stones; the carnelians as its fruits, its vines are too beautiful to look at, the leaves are of lapis lazuli, the grapes are too lush. Pure water runs through the garden, and right in the middle is the tree bearing papparduli stone."

The mystery of the crystal city finds a place to itself in the book of Enoch, too. In fact, we can say that in all ancient tablets, the houses and equipment of the Anunnaki were always described with the use of crystals and special stones. The Admiral, too, is carried, together with his radioman, on a platform with no wheels to the city center.

"It moves us toward the glowing city with great swiftness. As we approach, the city seems to be made of a crystal material.

Soon we arrive at a large building that is a type I have never seen before. It appears to be right out of the design board of Frank Lloyd Wright, or perhaps more correctly, out of a Buck Rogerssetting!! We are given some type of warm beverage, which tasted like nothing I have ever savored before. It is delicious. After about ten minutes, two of our wondrous appearing hosts come to our quarters and announce that I am to accompany them. I have no choice but to comply. I leave my radioman behind and we walk a short distance and enter into what seems to be an elevator.

We descend downward for some moments, the machine stops, and the door lifts silently upward! We then proceed down a long hallway that is lit by a rose-colored light that seems to be emanating from the very walls themselves! One of the beings motions for us to stop before a great door. Over the door is an inscription that I cannot read. The great door slides noiselessly open and I am beckoned to enter.

One of my hosts speaks.

> *'Have no fear, Admiral, you are to have an audience with the Master...' "*

The emittence of an old rose colored light through the walls while proceeding through a long corridor reminds us of the twenty-seven crystal stones in the crevice mentioned in the Lugal-e Text. These crevices, which we can even see today in the Great Pyramid's long and narrow corridor that is known as the "Gallery", served many purposes. (5)

Had the inscription above the gate been an undeciphered ancient language, or a newer one, the admiral would have found it in the later years for sure. Therefore, we can assume that it was a language that was not used in the world.

"I step inside and my eyes adjust to the beautiful coloration that seems to be filling the room completely.

Then I begin to see my surroundings. What greeted my eyes is the most beautiful sight of my entire existence. It is in fact too beautiful and wondrous to describe. It is exquisite and delicate. I do not think there exists a human term that can describe it in any detail with justice!

My thoughts are interrupted in a cordial manner by a warm rich voice of melodious quality,

> *'I bid you welcome to our domain, Admiral.'*

I see a man with delicate features and with the etching of years upon his face. He is seated at a long table. He motions me to sit down in one of the chairs."

Right away we think of Enki sitting on his throne...

"After I am seated, he places his fingertips together and smiles.

He speaks softly again, and conveys the following:

> *'We have let you enter here because you are of noble character and well-known on the Surface World, Admiral.'*

Surface World, I half-gasp under my breath!"

As you can see, it is being clearly mentioned here that the facility is underground. Also, the message as "we have the control" is being given. Let us remind that the soon-to-be writer Endubasar was met by Enki on a similar type of encounter as explained in the Lost Book of Enki by Zecheria Sitchin. (6)

" 'Yes,' the Master replies with a smile, 'you are in the domain of the Arianni, the Inner World of the Earth. We shall not long delay your mission, and you will be safely escorted back to the surface and for a distance beyond. But now, Admiral, I shall tell you why you have been summoned here."

Could the word Arianni be related to the city of Arcadia, and the Ari race in the animation movie Les Mondes Engloutis?

" 'Our interest rightly begins just after your race exploded the first atomic bombs over Hiroshima and Nagasaki, Japan. It was at that alarming time we sent our flying machines, the "Flugelrads", to your surface world to investigate what your race had done.' "

At this point, we can talk about the resemblance of the Flugelrad to the eagle-ships the Sumerian Deities utilize. These atomic bombs were, once again, on the agenda during the 1954 US-Anunnaki meeting, and the Anunnaki had prohibited the use of these bombs to the US.

> *"'That is, of course, past history now, my dear Admiral, but I must continue on.*
>
> *You see, we have never interfered before in your race's wars, and barbarity, but now we must, for you have learned to tamper with a certain power that is not for man, namely, that of atomic energy. Our emissaries have already delivered messages to the powers of your world, and yet they do not heed. Now you have been chosen to be witness here that our world does exist.*
>
> *You see, our Culture and Science is many thousands of years beyond your race, Admiral.'*

I interrupted,

> *'But what does this have to do with me, Sir?'"*

It seems like the person who speaks to the Admiral gives away the information that he is an Anunnaki, because the Sumerians, who had suddenly initiated a civilization without precedency at 3800 BCE, said "Whatever looks beautiful, we achieved with the grant of the Anunnaki." Human civilization was initiated by the Sumerians, but there was the Anunnaki civilization before that.

"The Master's eyes seemed to penetrate deeply into my mind, and after studying me for a few moments he replied,

> *'Your race has now reached the point of no return, for there are those among you who*

> *would destroy your very world rather than relinquish their power as they know it...'*

I nodded, and the Master continued,

> *'In 1945 and afterward, we tried to contact your race, but our efforts were met with hostility, our Flugelrads were fired upon.*
>
> *Yes, even pursued with malice and animosity by your fighter planes. So, now, I say to you, my son, there is a great storm gathering in your world, a black fury that will not spend itself for many years. There will be no answer in your arms, there will be no safety in your science.*
>
> *It may rage on until every flower of your culture is trampled, and all human things are leveled in vast chaos. Your recent war was only a prelude of what is yet to come for your race. We here see it more clearly with each hour... do you say I am mistaken?'*
>
> *'No,' I answer, 'it happened once before, the dark ages came and they lasted for more than five hundred years.' "*

We stop at this point, and ask how the dark ages had come to an end five hundred years ago. If we summarize all that has happened within the last five hundred years; the reform movements, the Renaissance movement, geographical discoveries, understanding the Solar System, deciphering the ancient languages, and the disclosure of a hidden history... We have already suggested that all this was the preparation for the Age of the Aquarius. Enki, who is the leader of the Age of the Aquarius, says that the age before him was the dark age, and adds that it is over.

" 'Yes, my son,' replied the Master, 'the dark ages that will come now for your race will cover the Earth like a pall, but I believe that some of your race will live through the storm, beyond that, I cannot say.

We see at a great distance a new world stirring from the ruins of your race, seeking its lost and legendary treasures, and they will be here, my son, safe in our keeping.' "

The person, who speaks to the Admiral, gives him the message as "Reach the ancient information". We think that he means disclosure of the real history through the light of the information on the tablets by "lost legendary treasures". He may be referring to the chosen ones, who will awaken and proceed towards the realities, which will be protected by Enki and his team, who will rule the world.

"'When that time arrives, we shall come forward again to help revive your culture and your race.

Perhaps, by then, you will have learned the futility of war and its strife...and after that time, certain of your culture and science will be returned for your race to begin anew. You, my son, are to return to the Surface World with this message...' "

These last words remind us of Noah's Deluge. In both the Epic of Gilgamesh and the Text of Atra Hasis, as well as various other ancient tablets, Enki chooses Noah (Utnapishtim, Ziusudra,) and gives messages similar those above. The unbelievable similarities between Enki's warnings and style and the Master's warnings and style cannot escape from our eyes.

We cannot help but ask whether a new deluge is to arrive, and this time, whether the "awakened ones" will be saved. However, once we look at the process since that day, the opinion that the salvation in mention is actually a "Harvest" depending on tests and election.

"With these closing words, our meeting seemed at an end. I stood for a moment as in a dream... but, yet, I knew this was reality, and for some strange reason I bowed slightly, either out of respect or humility, I do not know which.

Suddenly, I was again aware that the two beautiful hosts who had brought me here were again at my side.

> 'This way, Admiral,' motioned one.

I turned once more before leaving and looked back toward the Master. A gentle smile was etched on his delicate and ancient face.

> 'Farewell, my son,' he spoke, then he gestured with a lovely, slender hand a motion of peace and our meeting was truly ended.

Quickly, we walked back through the great door of the Master's chamber and once again entered into the elevator.

The door slid silently downward and we were at once going upward. One of my hosts spoke again, 'We must now make haste, Admiral, as the Master desires to delay you no longer on your scheduled timetable and you must return with his message to your race.'

I said nothing. All of this was almost beyond belief, and once again my thoughts were interrupted as we stopped. I entered the room and was again with my radioman. He had an anxious expression on his face.

As I approached, I said,

> 'It is all right, Howie, it is all right.'

The two beings motioned us toward the awaiting conveyance, we boarded, and soon arrived back at the aircraft. The engines were idling and we boarded immediately.

The whole atmosphere seemed charged now with a certain air of urgency. After the cargo door was closed the aircraft was immediately lifted by that unseen force until we reached an altitude of 2700 feet. Two of the aircraft were alongside for some distance guiding us on our return way.

I must state here, the airspeed indicator registered no reading, yet we were moving along at a very rapid rate.

22:15 Hours - A radio message comes through.

'We are leaving you now, Admiral, your controls are free. Auf Wiedersehen!!!!'

We watched for a moment as the flugelrads disappeared into the pale blue sky.

The aircraft suddenly felt as though caught in a sharp downdraft for a moment. We quickly recovered her control. We do not speak for some time, each man has his thoughts...

22:00 Hours- We are again over vast areas of ice and snow, and approximately 27 minutes from base camp. We radio them, they respond. We report all conditions normal....normal. Base camp expresses relief at our re-established contact.

23:00 Hours- We land smoothly at base camp. I have a mission.....

March 11, 1947

I have just attended a staff meeting at the Pentagon. I have stated fully my discovery and the message from the Master.

All is duly recorded. The President has been advised. I am now detained for several hours (six hours, thirty-nine minutes, to be exact.) I am interviewed intently by **Top Security Forces** *and a medical team. It was an ordeal!!!!*

I am placed under strict control via the national security provisions of this United States of America. I am ORDERED TO REMAIN SILENT IN REGARD TO ALL THAT I HAVE LEARNED, ON THE BEHALF OF HUMANITY!!! Incredible!

I am reminded that I am a military man and I must obey orders.

30/12/56 - FINAL ENTRY

These last few years elapsed since 1947 have not been kind...

I now make my final entry in this singular diary. In closing, I must state that I have faithfully kept this matter secret as directed all these years. It has been completely against my values of moral right. Now, I seem to sense the long night coming on and this secret will not die with me, but as all truth shall, it will triumph and so it shall.

This can be the only hope for mankind. I have seen the truth and it has quickened my spirit and has set me free! I have done my duty toward the monstrous military industrial complex.

Now, the long night begins to approach, but there shall be no end.

Just as the long night of the Arctic ends, the brilliant sunshine of Truth shall come again... and those who are of darkness shall fall in its Light... FOR I HAVE SEEN THAT LAND BEYOND THE POLE, THAT CENTER OF THE GREAT UNKNOWN.

Admiral Richard E. Byrd
United States Navy
24 December 1956"

After all this, the Admiral returns and gives the message to US at Pentagon. At the end, the Admiral, who is asked to remain silent, makes all this public three months prior to his death. Sure; we cannot claim that what Admiral Byrd experienced is completely real. However, once we look at history, we can say that similar encounters have been experienced quite often. Furthermore, we think that, during these current times, many people have contact with the Anunnaki through visions and dreams without knowing that they are the Anunnaki.

Since we know that the Age of Pisces belonging to Sin is about to come to an end, and that of Enki; the Age of Aquarius is about to begin, the messages above gain meaning. We can suggest that the Anunnaki have at least four or five other bases in other locations just like this one in the North.

Although they have not named them as Anunnaki, once, those who think like us have put forward the "Hollow Earth" thesis for these underground civilizations. There were hundreds of scientists defending the Hollow Earth Theory, who were supporters of the National Socialist Sciences (NSS), and each one was handling research for this science. In NSS, research was carried on fields like parapsychology, energy, space research, and past lives besides what the current science covers.

The thesis of Hans Horbirger and H. S. Bellamy, two prominent defenders of NSS, have been championed by millions even after the Nazi Germany. Cyril Henry Hoskin published the book "The Third Eye" in Britain in 1957 in support of Horbiger. Despite being British, the author of this book claims that he is actually a Tibetan Lama, and his name is Lobsang Rampa. In his book "The Rampa Story", Rampa tells in great detail how he exchanged bodies with a European with a broken soul in the astral plane. According to one claim, Rampa is one the Germans envoyed to Tibet by Hitler, and he stayed there until long after the War was over. The British newspapers and the intelligence service researched into the identity of Rampa, but they could not retrieve anything. There are two possibilities on the Rampa case: He was either a real lama as he claimed, or he was verbalizing the National thesis and those of Horbiger's by reflecting the information shared with him.

When he leaked the documents in 2013, Edward Snowden, who is a former CIA intelligence contractor, has stated that a race resembling to the human beings has their own underground civilization, and that they are looking at us as we look at the ants. The files, deemed ridiculous by many, have been leaked from NSA, and the copies have been sent to CIA. According to the leaked documents, those living underground are not aliens. However, they have been living underground for millions of years in a more advanced state than the humans. (7)

According the documents, this species, which has its own system, keeps the information at centers under the sea, and obstructs the human race to reach these systems. Another information that attracts attention in the files is that this species has been developing simultaneously with the Homo Sapiens, only reaching at a more advanced state. The files also include the information that this species protects itself with nuclear weapons, and that they have precautions against intruders, in case they are reached at.

Edward Snowden, who had to leave US following the leakage of the files, accepted the offer from Venezuela, and he left for Caracas. Besides, Snowden also leaked many UFO photographs and images creating a huge impact the world over.

Here is what we have to say as a conclusion: If the person who encountered with Admiral Byrd is Enki, who is the name behind science and communications, he has given a very clear message to us, which appeals to us in the current times as well; The Age of the Aquarius, lasting for 1260 years, will only be the age science. Those who can adopt to science will survive, and those who cannot will be lost in the dark ages forever...

Sources:

1. http://www.sechaber.com.tr/gecmisten-gunumuze-ulasan-cok-ozel-bir-mesaj/
2. http://www.anime.gen.tr/yazi.php?id=60
3. http://www.cnnturk.com/2012/bilim.teknoloji/bilim/02/10/sibiryada.canli.mamut.goruldu/648523.0/index.html
4. Enki ve Ninmah Miti, http://etcsl.orinst.ox.ac.uk/cgi-bin/etcsl.cgi?text=t.1.1.2&charenc=j#
5. Lugal-E, http://www.gatewaystobabylon.com/myths/texts/ninurta/exploitninurta.htm
6. SİTCHİN, Zecharia, Enki'nin Kayıp Kitabı, Ruh ve Madde Yayınları, İstanbul, 2008
7. http://www.sechaber.com.tr/bizi-yeraltindan-karinca-gozuyle-izleyen-medeniyet/

SECTION 6

MESSAGES FOR TODAY FROM THE BOOK OF ENOCH

The Book of Enoch Discovered in Ethiopia

This teaching presents an appraisal of the Earth for all humanity, for all my children who will live on Earth, and the next generations who will walk in the path of justice and peace. Let not your spirit be saddened by the negativities of the times, for the Holy One has appointed days for all things.

Enoch

The Book of Enoch: 2000s

The Book of Enoch, which we have cited from a few times while talking about the meeting of Admiral Byrd and Enki, presents very important information for the current days and for the future. We think that Enki has given multiple times more information to humanity in this book than what he gave to the Admiral. Although numerous research have been carried out abroad on the Book of Enoch, the first interpretation of its possible connection to the Anunnaki is being made hereby. For about two months, we have concentrated on this subject, and tried to the best of our capacity to decipher the codes, and reach the hidden information. The most challenging issue during this process was the different styles used in the book. During a detailed examination of the Book of Enoch, one comes across with at least four or five styles. Consequently, this makes us think that the book was written by different people at different times. At the end, we saw that the most important information was written using the style as below. Therefore, while deciphering the book, we emphasize the information, the energy and the coded narratives delivered through these or similar words.

The Book of Enoch begins with these statements:

These are the words of the blessing of Enoch. (It explains) how he blessed the 'Chosen' and the 'Righteous' at times of distress. Enoch, whose eyes were opened, and who saw a holy vision, said "That vision was shown to me by the angels. I heard everything from them, and I understood what I saw; what was being shown was not intended for this generation, but for a distant generation that belonged to the Chosen. Concerning the Chosen, I spoke with the Holy and the Great One; the God of Earth, who will come out his dwelling. (Book 1, Chapter 1)

We were already thinking that expressions like 'Chosen' and 'Righteous', which are used in this text, were adjectives defining the Anunnaki, and that they were being used to describe someone. In the Jewish mysticism, too, these were used as adjectives, and they were designated to certain societies. So, among the Anunnaki, who could be this "Chosen One", who is said to not have come yet? And again, who could be "The Righteous One" or the "The Holy and the Great One"? One thing that is for sure in this text is that the 'Holy and the Great One" is also the "God of Earth". More importantly, what was being shown to Enoch was not intended for the generation existing at the time the book was written, but rather for a more distant one. We wonder if the fact that the book was discovered at the end of the 18th Century, stands as a sign that the distant generation mentioned in the book might actually be our generation.

All these coded messages were getting even more interesting in the latter parts of the book. Therefore, all the solutions we found would have been incomplete without the key for the lock. As a result of our studies, we came to the conclusion that the keywords to unlock the lock, that is; "The Holy One" or "The Holy and the Great One" were actually adjectives of Enki. Read with this in mind, the consistent narratives in the further pages of the book were giving us in all clarity whatever was happening. What about Enlil? His adjective, we diagnosed as "The Ancient One".

Enki's time refers to the Age of Aquarius; the new age in front of us that will last for 2160 years. "Chosen One" and "Righteous One" are the adjectives of this age. "A distant generation" is intended for the last one degree between the Age of Pisces and the Age of Aquarius; in other words, our generation. All this means that the Book of Enoch gives messages right to this century. Once we accept the person who met with Admiral Byrd as Enki, we can also see what we have been preparing for during the last five hundred years. Enki, who defines the ages before him as the ages when evil abounded, and who says that all that was left in the past, tells us through coded narratives in the Book of Enoch what this age will bring to us. This book, which appeared when Enki's Age of Aquarius was just starting, has the purpose of giving information to the people "Invited" and "Chosen" by him.

In the upcoming sections, we will be explaining very clearly why some people are awakening in current times. Additionally, other than the chosen ones, the book refers to those among us, who have the dominant Anunnaki genes, as the "Holy Ones", and explains that they have different duties than the rest. The greatest prophecy of his book is that our world will be a place full of peace in the next period.

The Authenticity of the Book of Enoch

The original of the Book of Hanok or Enoch in Hebrew is lost, but it is definite that it once existed, because its pieces mixed with the Aramaic dialect have been discovered together with the Dead Sea Scrolls.

The seven copies of the Book of Enoch written in Aramaic, which were discovered in the fourth cave of the Qumran Caves, are dated to 2nd Century BCE corresponding to the end of the eras before Christianity. In these copies, there is information on the names of the leaders of the "Fallen Angels", their descend to the earth in order to choose wives to themselves, the birth of the giants, their imprecation, Enoch's arrival at the "Heaven of Righteousness" and about astronomy. The date when the Book of Enoch was written is unknown. However, research shows that the book was not written all at one time; it went through changes through time; and a few different writers had included additions in it.

In the Book of Genesis in the Torah, Hanok's/ Enoch's ascend to the skies have been recorded as part of the ancestors before the Deluge. Just like Enmeduranki in the Sumerian Kings List, Enoch is the seventh of the ten ancestors. In reference to this extraordinary experience, the Bible settles for only mentioning that Enoch, who was 365 years old at that time, disappeared as he was in the path of God, because God had chosen to take him to his side. Luckily, the Book of Enoch has reached us in two versions without getting lost in the course of thousands of years, and provides many more details than the Torah.

This piece of work, which has been translated into Greek and Latin, and which has been cited in many places, has been accepted as a holy text by almost all the authors of the New Testament, and due to this, it has been able to reach current times. Besides the work known as The First Book of Enoch, which was discovered by James Bruce in a Habesh cave in Ethiopia in 1773, there is also "The Book of the Secrets of Enoch", which was saved in the Slavonic language, and discovered in a monastery in Russia recently. The text we have tried to interpret is that one of the First Book of Enoch, and the Turkish translation of it has been made by Gunyuz Keskin; Hermes Publishing House. Nevertheless, it is hard to tell how much of the books compiled just before the Christian Era started, were ancient or imaginative and speculative.

Based on the information revealed in these sources, Enoch did not take one, but two celestial journeys. On the first one, he was taught the "Secrets of the Heavens", and asked to pass this information to his sons. Yet, the second journey is only a departure. That is why there is mention of "Enoch never returned from this journey." and "He disappeared, because Elohim took him by his side." In the Book of Enoch, Elohim is portrayed as a member of the team of angels, who fulfills the works ordered by God. But, for us, who believe in the Ancient Astronauts Theory, Elohim is one of the Anunnaki, who came from Nibiru. Consequently, Enoch did not die, but rather, his lifespan got lengthened, and his frequency heightened; thus, he was taken by the Anunnaki.

The Book of Enoch tells that the prophecies were written firsthand and in person by Enoch. Despite all the academic opinions defending the compilation of the book right before the Christian era, the fact that books dated to earlier times have citations from it, and that it has parallelism with texts other than the Holy Book, makes it definite that this text is based on ancient texts.

Enoch, who has been provided with the information on the Heavens, the Earth and their mysteries, has been told to put the prophecies of the future in writing. According to the Book of Jubilees, Enoch was shown "what had happened and what will happen", and he put the secrets of the creation and the phases of the events on Earth in writing for the next generations. God has placed a "map" on Earth, and determined the destiny of the planet and everything on it. The texts surviving from the times prior to the Deluge are about "the starts, the mids and the ends".

Now, let us start examining the parts that we deem as the most important. (1)

Enoch Says: This wisdom teaching is an appraisal of Earth for the next generations

This wisdom teaching has been written by Enoch. This teaching presents an appraisal of the Earth for all humanity, for all my children who will live on Earth, and the next generations who will walk in the path of justice and peace. Let not your spirit be saddened by the negativities of the times, for the Holy One has appointed days for all things. The righteous man will rise from sleep, and will walk in the path of righteousness, and all his paths, and his journeys, will be in eternal goodness and mercy. He will show mercy to the loyal ones and give them eternal justice and power. He will walk in eternal light. And sin will be destroyed in darkness, forever, and from that day on will never be seen again. (Book 4, Chapter 91)

Through these lines, Enoch calls especially to the awakened and the ones who have awareness, and gives them advice as "Let not your spirit be saddened by the negativities of the times." "The Holy One (Enki) has appointed days for all things." refers to the peace to come during the Age of the Aquarius.

Wisdom, Light, Joy, Peace

But, for the chosen; there will be light, joy, and peace. They will inherit the earth. But for you, the impious, there will be no salvation. You will all be surrounded by a curse. Wisdom will be given to the chosen. They will all live, and will never again sin, either through injustice, or through pride. In all the days of their lives, they will never be convicted, they will never be strayed from the path, they will not sin throughout their lives, and they will not die of divine wrath or anger. But they will complete the number of the days of their life. Yes, they will reproduce in peace, and their joyful years will turn into an eternal happiness and peace. (Book 1, Chapter 6)

If "Chosen" refers to the people whom we consider as the awakened people of our current times (and we think it does), it is said that what awaits them is light, joy and peace. Looking at the lives of the people who consider themselves as awakened, we can ask these questions: How often they use the word "light"? How much room do joy and peace have in their lives? When did the awareness begin, and how did wisdom make these people different than the rest? Why did they find themselves chasing the truth at all times? Although there are so many activities to steal the time on earth, why do they read and research so much, and continuously question the existence? Right, can they easily sin? How much value they place on human rights, animal rights and children's rights? Can they do the activities which seem to be inherently wrong for them?

When you give the answers to these questions, if you notice that you are like one of the people Enoch mentions, you may be among the "chosen" as well.

Enoch in Enlil's Spacecraft

And a vision appeared to me: - Behold; clouds called me in, and I was pulled into the mist. And the path of the stars, and flashes of lightning, hastened me and drove me. And in the vision winds caused me to fly, and hastened me, and lifted me up into the sky. And I proceeded ascending until I came near a wall, which was made of crystal stones, and surrounded by a tongue of fire glowing with heat. I started feeling afraid. And I went into the tongue of fire, and came near to a large house, which was built of crystal stones. The foundation of the house was made of crystal stones, too. Its roof was like shining stars and flashes of lightning, and among them was fiery cherubim. The sky was clear as water. There was a fire burning around its wall and its door was ablaze with fire. I went into that house. It was as hot as fire and as cold as snow. There was neither pleasure nor life in it. Fear covered me and trembling took hold of me. (Book 1, Chapter 14)

We can say that Enoch's descriptions are quite on the point as someone who did not know the technology of our times. In fact, we feel like we are on the film set of a sci-fi movie. Ascending to the skies in a vehicle covered in a mist camouflage; entering the main spacecraft in a vehicle generating winds like a helicopter; a spacecraft which has a protection shield of hot-red fames; and the spotlights in the ceiling and the mosaics on the floors of the spacecraft constructed with crystals, all make sense to us.

The Torah gives similar information on the spacecraft appearing through the clouds. According to the narrative, *the Lord had guided them during the day hiding in a pillar of clouds, and he had provided light at night with a pillar of fire so that the Israelites could travel by day or by night.* (Exodus 13:21/22) *In another chapter, they had seen the awesome glory of the Lord in the cloud, when they looked out toward the desert.* (Exodus 16:10).

The glory of the Lord not only appeared amongst the clouds, but also landed on Earth.

When Moses went up to the mountain, the cloud covered the mountain. And the glory of the Lord settled down on Mount Sinai, and the cloud covered it for six days. On the seventh day the Lord called to Moses from inside the cloud. To the Israelites at the foot of the mountain, the glory of the Lord appeared at the summit like a consuming fire. (Exodus 24:14/17)

Furthermore, the details of this descend were provided. First the Lord had told Moses "*I am going to come to you in a dense cloud*" (Exodus 19:9) and then, he had landed on Earth in front of a crowd exceeding a million.

Exodus 19: 16 On the morning of the third day, thunder roared and lightning flashed, and a dense cloud came down on the mountain. There was a long, loud blast from a ram's horn, and all the people trembled in the camp.

Exodus 19: 17 Moses led them out from the camp to meet with God, and they stood at the foot of the mountain.

Exodus 19: 18 All of Mount Sinai was covered with smoke because the LORD had descended on it in the form of fire. The smoke billowed into the sky like smoke from a brick kiln, and the whole mountain shook violently.

Exodus 19: 19 As the blast of the ram's horn grew louder and louder, Moses spoke, and God thundered his reply.

Exodus 19: 20 The LORD came down on the top of Mount Sinai and called Moses to the top of the mountain. So Moses climbed the mountain.

Exodus 19: 21 Then the LORD told Moses, "Go back down and warn the people not to break through the boundaries to see the LORD, or they will die.

We see the winds, another feature of Enoch's description, on Elijah's being taken into the skies: *As they were walking along and talking, suddenly a chariot of fire appeared, drawn by horses of fire. It drove between the two men, separating them, and Elijah was carried by a whirlwind into heaven.* (2 Kings 2:11).

When we look into the explanations here, we see that the Lord, who speaks in thunders, uses a technology like that of a system of speakers, and that the spacecraft either creates fumes or uses fumes as camouflage. We also notice interesting details like the dense heat and brightness emitted from the spacecraft's ignition system creating fear in everyone, and the announcement saying whoever approaches the bottom part of the spacecraft will die. Right, so, how was this spacecraft, which was transcripted as the Lord's Glory, shaped? The Lord himself gives the answer to this:

You have seen what I did to the Egyptians. You know how I carried you on eagles' wings and brought you to myself. (Exodus 19:4)

The same Lord causes the death of a hundred and eighty-five thousand people during the Assyrian camp siege in Jerusalem in 701 BCE, after hearing King Hezekiah's prayers and by sending fire from the Heavens. This incident is mentioned at three separate points in the Bible. (Isaiah 37:36, 2 Chronicles 32:21, 2 Kings 19:35)

Hezekiah witnesses this incident with his bare eyes. The eagle wings on his 2700 years old seal do not escape from our attention. The eagle wings is a nice depiction of the Eagle Ships drawn in Sumer. (Photo 1)

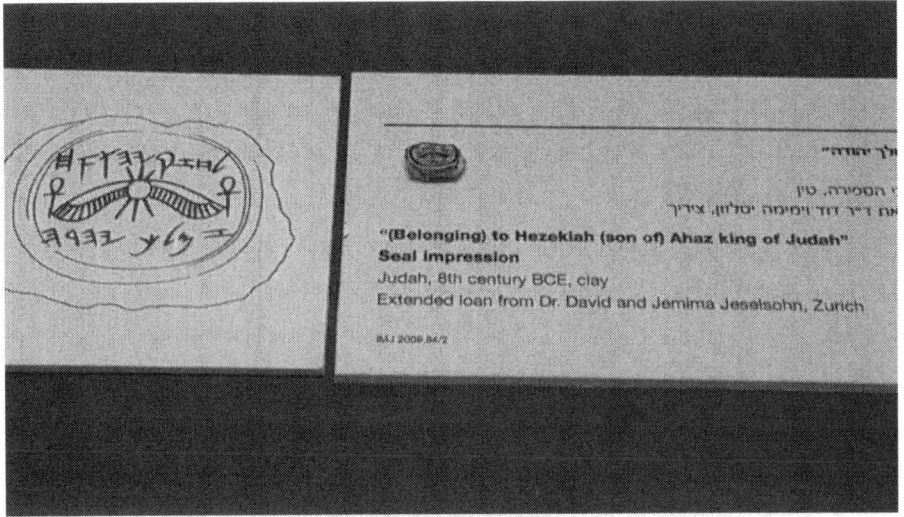

Photo 1: Hezekiah's Seal

We say that the leading role in all these incidents belongs to a technologically superior Anunnaki, not the Lord, and we continue with the Book of Enoch:

And as I was shaking and trembling, I fell on my face. I saw another vision. There was a second house which was larger than the former. It was completely built of fire, and all its doors were open before me. It was excellent in everything, and was very big. It so excelled in glory and splendor and size, so that I am unable to describe to you its glory and its size. Its floor was of fire. Above, there was lightning and stars. And, its roof also was a burning fire. On a more careful glance, I saw in it, a high throne, and its appearance was like crystallized, and its surroundings like the shining Sun. There was the sound of cherubim. From underneath the throne there flowed out rivers of fire so that it was impossible to look at it. He who is Great in Glory sat upon it, and his raiment was brighter than the Sun, and whiter than any snow. Due to his grandeur, no Angel could enter at the appearance of the face of Him. And, no human being could look at him, either. A sea of fire burnt around Him, (Book 1, Chapter 14)

Most probably, Enoch enters the cockpit of the spacecraft and meets the owner of the ship. Obviously, he is too impressed by the lighting, and he keeps on emphasizing it. He describes the light as "rivers of fire". Let's see who the owner of the ship is, and what he wants.

The Lord told me: "Do not be afraid, Enoch, you, the righteous man, and scribe of righteousness. Come here and hear my voice. Go say to the Watchers of Heaven, who sent you to petition on their behalf: You ought to petition on behalf of men, not men on behalf of you. Why have you left the High, Holy and Eternal Heaven, and lain with women, and become unclean with the daughters of men, and taken wives for yourselves, and done as the sons of the earth, and begotten giant sons? Why did you become unclean upon the women's blood while you were leading a spiritual, holy, and an eternal life? Why did you mix with human blood? Why did you begot children through the blood of flesh, and lusted after the blood of men, and produced flesh and blood, as they do, who die and are destroyed. (Book 1, Chapter 15)

Hereby, the figure who speaks to Enoch reveals who he is: He is no other person but Enlil. As we have told in our book "Amon Ra", the Igigi's (Watchers) choice of wives for themselves among the humans had disturbed Enlil very much. This disturbance had never ceased, because of the marriages between the Anunnaki and the human beings. It had even caused very harsh decisions to be taken during the Deluge. As far as we can follow through the text above, the Igigi sent Enoch, who was favored very much by Enlil, as an intermediary. Enlil answers him in a humorous manner:

"Go say to the Watchers of Heaven, who sent you to petition on their behalf: You ought to petition on behalf of men, not men on behalf of you."

"Giant sons" refers to the human beings as we know as half-deities. In our previous books, we had suggested that the Anunnaki were about two and a half to three meters tall. Through their marriages with humans, two meter tall children are born, and they are called as giants...

A volcano in the Mediterranean

And they took me to the Water of Life, as it is called, and to the Fire of the West, which receives every setting of the Sun. (Book 1, Chapter 17)

"The Great Sea" is a term mentioned in the Bible quite often, and it refers to the Mediterranean. Hereby, we can easily say that Enoch has seen a volcano in the Mediterranean (perhaps Mount Etna).

The Hollow Earth Theory and the Country in the North

And from there I went to another place, towards the west, to the ends of the Earth. And I saw a fire that burnt and ran, without resting or ceasing from running, by day or by night, but continued in exactly the same way. And I asked saying: "What is this which has no rest?" Then Raguel, one of the Holy Angels, who was with me, answered me, and said to me: "This burning fire, whose course you saw towards the west, is the fire of all the Lights of Heaven." (Book 1, Chapter 23)

Hereby, the light in mention is an artificial light, and Enoch adds that it is at the end of Earth. We think that the source of light in mention is the place Admiral Byrd had talked about. This is what he had said:

"1005 Hours- I alter altitude to 1400 feet and execute a sharp left turn to better examine the valley below. It is green with either moss or a type of tight knit grass. The Light here seems different. I cannot see the Sun anymore."

Enoch goes to the Garden of Aden over the Sea of Eritrea

And after these fragrances, to the north, as I looked over the mountains, I saw seven mountains full of fine nard, and fragrant trees of cinnamon and pepper. And from there, I went over the summits of those mountains, far away to the east, and I went over the Eritrean Sea, and I was far from it, and I went over the Angel Zotiel. And I came to the Garden of Righteousness, and I saw beyond those trees many large trees growing there, sweet smelling, large, very beautiful and glorious, (Book 2, Chapter 31)

Once we reach the further sections of the text, we understand that "The Garden of Righteousness" mentioned in the text is Aden. The interesting point about it all is that Enoch goes to the Garden of Righteousness by "flying", in other words, on a spacecraft, and that he describes his route passing over the Sea of Eritrea. In this case, where could this Aden be? Since we know that Enlil lives in Aden, we are left with two alternatives: One is Enlil's city Nippur, and the other is Enlil's base Baalbek.

Enoch in Machu Picchu

And from there I went towards the west, to the ends of the Earth, and I saw there three open windows. There were windows and outlets as I saw in the east. (Book 1, Chapter 34 – Journey to West)

We think that Enoch points out to Machu Picchu with these words. Although the history of Machu Picchu is not known definitely, it has been suggested that the city was the training and disciplinary place for the Incas. However, this theory has been discredited nowadays. According to Hiram Bingham from the Yale University, who discovered Machu Picchu, and who made research on it, it is actually Tampu-tocco, the legendary city of the Incas. (2)

What makes one believe that Machu Picchu is actually the legendary Tampu-tocco is the hint on the Three Windows. In the book "Peru's Historic Ancient Memories" written by Fernando de Montesinos in 1644, there is this record: "Inca Roca, the King of the Incas, ordered a wall with three windows, which was the symbol of the family of his fathers he was descendant of, to be built in the place he was born (Tampu-tocco)." The name of place where the royal family, who had left Cusco because of the catastrophe, went to means "Temple of the Three Windows". It may seem peculiar to us that a place becomes recognizable through its windows, but keep in mind that there are no windows in the houses in Cusco. The number of the windows of a place, in other words; its becoming recognizable by three windows, can only be the result of the uniqueness, archaism or holiness of such a structure that is really existing.

Just like all the other Titicaca structures finding a place in all the creation mythologies of the American nations, this city was also founded by the deities following the Deluge. We had defended that the Deluge had happened around 11 thousand BCE, and that the Anunnaki had arrived in South America after the Deluge, and had established these cities. If all these suggestions are correct, then, Machu Picchu is at least thirteen thousand years old, and the most important part of this city is what is known as the "Temple of the Three Windows" (Photo 2)

Photo 2: Temple of the Three Windows in Machu Picchu and Leonardo Da Vinci's The Last Supper painting

And the Lord will shut up those Angels, who showed iniquity, in that burning valley, which my great-grandfather Enoch had shown to me previously, in the west, near the mountains of gold and silver and iron and soft metal and tin. (Book 2, Chapter 66)

The place in mention hereby could be the ancient mining cities by and near Lake Titicaca, which is very near to Machu Picchu.

The Gates of Heavens; Is it the secret of going into space?

And from there I went to the ends of the earth, and I saw there large creatures, each different from the other, and also birds, which differed in form, beauty, and call - each different from the other. And to the east of those creatures, I saw the ends of the Earth, on which Heaven rests, and the open Gates of Heaven. And I saw the movements of the stars of Heaven, and counted the Gates out of which they come, and wrote down all their outlets. And I noted down their names, according to their constellations, their positions, their times, and their months, as the Angel Uriel, who was with me, showed me. (Book 2, Chapter 32)

And from there I went towards the north, to the ends of the Earth, and there I saw a great and glorious wonder at the ends of the whole Earth. There I saw three Gates of Heaven; through each of them north winds go out; when they blow there is cold, hail, hoarfrost, snow, fog, and rain. And from one Gate, it blows for benevolence; but when they blow through the other two Gates, it is with force, and, yes, it brings torment and trouble over the earth, and they blow with force. (Book 2, Chapter 33)

And from there I went towards the south, to the ends of the Earth, and there I saw three Gates of Heaven open; and the mist, and the rain, and wind, come out from there. And from there I went towards the east of the ends of Heaven, and there I saw the three eastern Gates of Heaven open, and above them, there were smaller Gates. Through each of these smaller Gates, the stars of Heaven pass, and go towards the west, on the path that has been shown to them. . (Book 2, Chapter 35)

The study carried out by NASA which found the hidden portals in Earth's magnetic field gives us very riveting information. (3) Based on the research of David Sibeck, who is the scientist on the THEMIS Project, from the Goddard Space Flight Center, energetic particles can pass through the openings formed by the portals, and heat the upper layers of the Earth's atmosphere, trigger the geomagnetic storms, and create the northern lights. In short, the particles transfer magnetic field from the Sun to the Earth through the X-points. According to DR. Sibeck, it is clear that there is a shortcut path between the Earth and the Sun, and this discovery can initiate the era of journey into the space. This finding can fundamentally change our understanding of the solar winds and magnetosphere interaction.

We say that the portals between the Earth and the Sun can minimize the amount of energy used while going into the space, and make journeys into the space much easier, and ask: Could there be similar shortcut paths between the other planets or stars and us? If there are, could these portals not be Enoch's Gates of Heavens?

The holy race to descend from the Heavens pretty soon: The Anunnaki

And it will come to pass in these days that the chosen and the holy race will come down from the high Heavens and their offspring will become one with the sons of men. In those days Enoch received books of indignation and anger and books of tumult and confusion. (Book 2, Chapter 39)

For this section, the translator has a very interesting choice of expressions. He or she says that the use of "holy race" in here is a complete mystery; that this race has nothing to do with the falling angels; and that a different race, who have seeded the mankind, is being referred to here. She or he has also added that it is impossible to place it in the mysticism of creation. According to us, the term "holy race" has been used for the Anunnaki.

Once we assume that the Book of Enoch is written for Enki and the Age of Aquarius, we can see what the increasing UFO traffic after the Rosewell accident, which is considered to be a milestone, is actually preparing us for. Through films, television series, and books, we are being prepared for an incredible encounter for the last one hundred years; to meet the "Holy Race", the Anunnaki. Considering that the earliest contacts were in 1947 with Admiral Byrd, and in 1954 with Eisenhower, we think that the timing of the meeting of the Anunnaki with the entire humanity cannot be too far. I personally think that this meeting will happen before 2060.

It is possible that the "Chosen", who are to descend together with the Holy Race, are the humans who have mingled with the Anunnaki after their lifespans were lengthened. In our book "The Last Call, Contact with the Anunnaki", we had explained in detail the people who were taken into the skies without dying. You must have heard about the stories of Enoch, prophets like Elijah, and some saints not dying, but rather going away by flying. These are the people Enoch calls as "Chosen".

Enoch in the mineral depots of the Anunnaki

And after those days, in that place where I had seen all the visions of that which is secret, I had been carried off by a whirlwind, and they had brought me to the west. There my eyes saw all the secrets of Heaven, and everything that exists on Earth: a mountain of iron, a mountain of copper, a mountain of silver, a mountain of gold, a mountain of nickel, and a mountain of lead. (Book 2, Chapter 51)

If you remember, Elijah was taken to the heavens in a whirlwind, too. The vehicles of the skies used a technology in those days similar to that of the helicopters, which could create powerful movements of air; and people made analogies to whirlwinds or hurricanes in order to be able to tell about it. In this part of the text, once again, Enoch is taken into a vehicle, and he goes to the West. He sees the mineral depots of the Anunnaki there, and he mentions them consecutively as "a mountain of iron, a mountain of copper, a mountain of silver, a mountain of gold, a mountain of nickel, and a mountain of lead" based on their amounts of stocks. Where this region lies is yet another mystery. However, if we need to make a guess, we can point to the ancient structures around Lake Titicaca.

Enlil's confession

And after this, the Ancient One repented, and said: "I have destroyed to no purpose all those who dwell upon the Earth." And he swore by His Great Name: "From now on I will not act like this towards all those who dwell upon the Earth. And I will put a signature in Heaven, and it will be a pledge of faith between us forever, so long as Heaven is above the Earth. And this will be in accordance with my command." (Book 2, Chapter 54)

We can understand from all the narratives that the "Ancient One" is Enlil. He had been in conflict with the "Igigi", the fallen angels, and he had obstructed the announcement of the deluge to come to the humans. Besides, as we keep on repeating constantly, since the day the Anunnaki arrived in our world, the leader has been Enlil, and he has passed the leadership on to Marduk at the starting point of the Age of the Ram. In this respect, we can say that the title of "Ancient One" fits so well to Enlil, who has ruled the World for 400 thousand years. Enlil's remorse after the Deluge was over is mentioned in other ancient tablets, too. In the eleventh tablet of the Epic of Gilgamesh, Gilgamesh meets with Utnapishtim (Noah), and Utnapishim tells about the entire period of the deluge. One of the things he tells is that he takes Utnapishtim and his wife Emzara on board the ship, following the remorse he feels, and that he lengthens their lifespans, and takes them among the Anunnaki. (4)

For us, the fundamental question is: What is the signature he puts in Heaven? We can find a hint on this in the Book of Creation. The Book of Creation or Genesis is the first book of the Torah, which is the first book of the first five books of the Tanakh and the Old Testament, and it has been accepted by experts that it takes Sumer as a source. In the text, a speech is given to Noah and his sons, who survived the deluge: (5)

"Yes, I am confirming my covenant with you. Never again will floodwaters kill all living creatures; never again will a flood destroy the earth." Then God said, "I am giving you a sign of my covenant with you and with all living creatures, for all generations to come. I have placed my rainbow in the clouds. It is the sign of my covenant with you and with all the earth. When I send clouds over the earth, the rainbow will appear in the clouds, and I will remember my covenant with you and with all living creatures. Never again will the floodwaters destroy all life. When I see the rainbow in the clouds, I will remember the eternal covenant between God and every living creature on earth."
(Genesis 9:11-16)

Blessed are you, the Holy and the Chosen, for your faith will be glorious!

And I began to speak The Third Parable.

Blessed are you, the Holy and the Chosen, for your faith will be glorious!

And the Holy will be in the light of the Sun and the Chosen in the light of eternal life. And there will be no end to the days of their life and the days of the Holy will be without number. And they will seek the light and will find righteousness with the Lord of Spirits.

The Holy will find peace with the name of the eternal Lord!

And after this it will be said to the Holy: "Seek the secrets of righteousness, the heritage of faith; for the world shines as the Sun, and darkness has passed away." And there will be a ceaseless light, and the days will never come to an end. The light will be powerful in front of the Lord of Spirits, and the light of uprightness will endure in front of the Lord of Spirits, forever.
(Book 2, Chapter 56)

Enoch says "Blessed are you, for your faith will be glorious!" to "the Holy", who have the Anunnaki genes, and "the Chosen", who point out to the awakened people. In addition to this piece of good news, we can say that the awakened people are protected. "And there will be no end to the days of their life" might be referring to the extension of the lifespans of the chosen. And, the part about "seeking the light" is very interesting. The fact that we hear the word "light" from every awakening person's mouth nowadays might be the proof of "seeking for the light".

And, the "eternal Lord" mentioned in the fourth article; we think that it is the greatest energy of the universe, which we know as "the Creator or the Maker".

Still another interesting feature is the holy in the heavens. These people, who come from the Anunnaki race; who are on special duty on Earth, and who guide us, have a permission to go into space. We are guessing that they have the freedom to go easily to Phobos and the bases on the Moon.

Those Who Do Not Sleep in Heavens

On that day, they will raise one voice, and will bless, praise, glorify, and exalt, in the spirit of faith, in the spirit of wisdom and of patience, in the spirit of mercy, in the spirit of justice and of peace, and in the spirit of goodness. Yes, they will all say with one voice: "Blessed is He, and blessed be the name of the Lord of Spirits for ever and ever." All Those Who Do Not Sleep in Heaven above will bless Him. All His Holy Ones who are in Heaven, will bless Him. All the chosen ones who dwell in the Garden of Life, and every spirit of light able to bless, praise and exalt, and hallow your Holy Name, and all flesh to the limit of its power, will praise, and bless, your holy name forever and ever. For great is the mercy of the Lord of Spirits, he has revealed all he did and created to the righteous and the chosen. (Book 2, Chapter 60)

What we consider important in this part of the text is the expression "Those who do not sleep in heavens". It may be referring to the existence of a technological system, which monitors the Earth, and the inhabitants of the Earth; and people, whom Enoch calls the "Holy", and whom we call the "Guides", as the managers of this system.

Although it is not a one-on-one analogy; perhaps, we can consider the Holy as the staff working for the "Heavenly customer relations", and there may be a designated awakened human being for each. They monitor us through the computers in front of them, and they give us messages, dreams, visions and intuitions when necessary.

The television series "Person of Interest", which is being advertised with the tagline "Cameras are everywhere! They are watching, listening, and recording. They know everything about you!" presents very valuable information on getting to know a system that is so similar to this one. (6)

In the meantime, it is specifically mentioned that the "Holy" are in the Heavens. In this case, two alternatives come to mind: They either live on Anunnaki bases such as the Moon and Phobos, or they are in spacecrafts.

On the other hand, there is a hint about the "Chosen": "All the chosen ones who dwell in the Garden of Life". If Garden of Life refers to the life on Earth, then, we can concur that the "Chosen" ones are the awakened people. The chosen ones, who have exceeded the limits of the natural, biological lives; who have not died, and whose lifespans have been extended, might be living on the bases on Earth.

Ancestors pre-dating History; Alexander and Enoch

Then, the name of the Son of Man was lifted from the
level of those who dwell upon the Earth to the level of the Lord of

Spirits. Yes, he was lifted on the chariots of the spirit to the Heavens, and his name vanished from among the humans. From that day on, I was not counted among them, and He placed me between the north and the west winds, where the Angels took the cords to measure for me the place for the chosen and the righteous.

And there I saw the first fathers of the people, who dwelt in that place since the beginning of the world, and the holy. (Book 2, Chapter 69)

Enoch "being ascended to the Heavens" is explicitly told in this section. In fact, in history, we see several examples of the holy and the chosen being taken among the Anunnaki, in ways similar to this. Being lifted to the level of the Lord of Spirits from among those who dwell on Earth is perhaps a change of frequency. As is known, the area we can see and hear is very limited. And, the Anunnaki are in frequencies outside this one, and they may be increasing the frequencies of the people they choose. Perhaps, the process being applied to Enoch is this. Enoch might have come to an end with his existence in the area we can see, and he might have ascended to the frequency level where the other chosen exist. In the place he goes to, there are other people, who are chosen just like himself.

Let us explain with an example: If someone is chosen, when the time comes, we will consider him dead, and think that we are putting him or her in the soil in front of the eyes of hundreds. It is a simple task for the current technology to implement this delusion into hundreds of people, all at one time. In the meantime, the "deceased" will be watching his or her own funeral ceremony, and he or she will be taken from among us. As soon as he or she is taken, all the junk genes in his or her body will be activated, and all the blockages cleared. His or her DNA software will be updated, and he or she will go back to the age thirties. Then, he or she will be taken to a place where there are other chosen ones like himself or herself. He or she will come cross with Ataturk right in front of him or her, hug him with love, and burst into tears. Afterwards, he or she will see Mozart, and get to know the other chosen ones, whom he or she admires. Then, the system will be explained to him or her, and he or she will be trained in his or her new duty, which is to guide the people. Considering that all the people he or she knows on Earth will be dead in one hundred years from now, this person will, perhaps, remain without getting old for thousands of years, and give guiding services to the next generations.

Alexander the Great had started a journey all the way to Egypt when he learned through the prophecies of Delphi that he was the son of Amon Ra, and when had this confirmed with his mother. He established his great empire during this journey, which he had started in order ask for a longer life from Amon Ra, and then, he died at a young age in his other dwelling, Babylonia. Maybe he did not die, and he was taken into the Heavens, too. Who knows; maybe he continued his journey as Zulqarnayn. When we read about his life, and see the adventures he goes through, we can understand that this is not such a freakish idea. (7)

One of the many state historians, who have recorded his life, is the Greek historian Callisthenes of Olynthus. He was assigned by Alexander to record the events, the victories and the adventures during the campaign to Asia. However, because of making him angry, he was put into jail, where he died, and his records disappeared mysteriously. Centuries after, a text in Latin, which was claimed to be the Latin translation of Callisthenes's original writings, was being circulated in Europe. Scholars refer to this text as "pseudo Callisthenes". This book is a magnum opus on the experiences of Alexander. Alexander has many history recorders, and in this way, his experiences can be examined from different angles.

In the account of one of Alexander's adventures, the historians mention that he visited a Nubian queen called Candace, whose beauty cannot be praised enough by any human, in the

South of Egypt. His accompaniments did not know about this. Yet, in fact, Alexander was not looking for love, but rather the secret of immortality. Following a period of joy together, as a present of departation, the queen agrees to give away the secret to a wonderful cave, where the deities gather.

Following the instructions, Alexander finds the holy place, and enters inside accompanied by a few soldiers, and sees a brightness similar to that of a star. The ceilings shine as if decorated with stars, just like in the description Enoch gives to Enlil about the spacecraft. The appearances of the deities resemble to human beings. Alexander, initially, gets afraid and puzzled. He sees figures lying down with eyes shining like light beams. But, he remains in order to see what will happen. One of these figures, who tells Alexander to stop there, asks: "Glad greetings, Alexander! Do you know who I am?" Alexander answers: "No, my Lord!" The other says: "I am Sesonchusis, the world-conquering king, who has joined the ranks of Gods." Despite having met with the very person he was searching for, Alexander was far from being surprised. Apparently, his arrival was expected; Alexander was invited in to the inner room where the Gods resided. He goes in, and sees God Serapis (in the version in Greek, this god is God Dionysus), seated on a throne, in a fire-bright lucidity. Alexander sees his chance to bring up the matter of his longevity, and asks: "How many years am I going to live?", but there is no answer from the god.

Then, Sesonchusis seeks to console Alexander, for the god's silence speaks for itself. "Though myself have joined the ranks of gods," Sesonchusis said, "I was not as fortunate as you. Although I have conquered the whole World, and subjugated so many peoples, nobody remembers my name. But, you have great fame... You will have an immortal name after death." In this way, he consoles Alexander. "You shall live after dying, therefore, you will not be dying."

The recorders interpret this part as Alexander to be immortalized by a lasting reputation. However, perhaps,

Alexander was being told that he would be taken into the Heavens, and that his lifespan would be lengthened, that is; the long lifespan mentioned in Enoch's Book.

A disappointed Alexander leaves the cave, and continues with his journey in the search for advice from other sages, a way to escape his mortal destiny, and the path to follow in order to join the others, who had managed to join the immortal deities before him.

Hereby, could the gods' cave be a spacecraft? Could the ancient ancestors there be the "Chosen" people taken into the Heavens once upon a time?

According to one version, among those whom Alexander was searching for, and got to know was Enoch, a biblical ancestor from the times before the deluge, who was the great-grandfather of Noah.

It is a place of mountains, "where Paradise, which is the Land of the Living, is situated," the "abode where the saints dwell." Atop a mountain there is a glittering structure, from which there extends skyward a huge stairway, made of 2,500 golden steps. In a vast hall or cavern Alexander sees "golden figures, each standing in its own niche," a golden altar, and two huge "candlesticks" measuring some twenty meters in height. Upon a couch nearby reclines the form of a man who was draped in a coverlet inlaid with gold and precious stones, and above it, worked in gold, are branches of a vine, having its cluster of grapes formed of jewels. The man suddenly wakes up and speaks, identifying himself as Enoch. "Do not pry into the mysteries of God," the voice warns Alexander. Heeding the warning, Alexander leaves to rejoin his troops; but not before receiving as a parting gift a bunch of grapes that miraculously are sufficient to feed his whole army.

In yet another version, Alexander encounters not one but two men from the past: Enoch, and the Prophet Elijah. The incident happens when Alexander is traversing an uninhabited desert. Suddenly, his horse is seized by a "spirit" which carries both the horse and its rider aloft, bringing Alexander to a glittering tabernacle. Inside, he sees two men. Their faces are bright, their teeth whiter than milk, their eyes shines brighter than the morning star; they are "lofty of stature, of gracious look."

Telling him who they are, they say that "God hides them from death." They tell him that the place is "The City of the Storehouse of Life," from where the "Bright Waters of Life" emanates. But before Alexander can find out more, or drink of the "Waters of Life," a "chariot of fire" snatches him away—and he finds himself back with his troops.

The important point here is that Elijah had not died, but lifted into Heavens according to the Bible, just like Enoch.

The Chariots in the Skies: UFOs

They are the conductors of the days and nights: the Sun, the Moon, the stars, and all the ministers of heaven, which make their circuit with all the chariots of heaven. (Book 4, Chapter 74)

Hereby, reference is made to celestial vehicles, in the modern sense; UFOs, which can go to the Sun, the Moon and the other celestial bodies in our Solar System...

I saw likewise the chariots of heaven, running in the world above to those gates in which the stars turn, which never set. One of these is greater than all, which goes round the whole world. (Book 4, Chapter 74)

Enoch, who says that one chariot is bigger than the others, might be referring to a main ship.

Enoch reads the Akashic Records

And, Uriel said, "O Enoch, look on the book which heaven has gradually dropped down; and, reading that which is written in it, understand every part of it." Then, I looked on all which was written, and understood all, reading the book "History of the Mankind" and everything written in it, all the works of man; and of all the children of flesh upon earth, during the generations of the world. (Book 4, Section 80)

I guess we would all like to read the book "History of the Mankind" which includes all the records of all human beings on Earth. We think that the main computer, which has artificial intelligence, designed by the Anunnaki has such records.

The joy taken from this wisdom will surpass the joy deemed worthy of eating some delicious food!

Now, my son Methuselah, all these things I speak unto thee, and write for thee! To thee I have revealed all, and have given thee books of everything. Preserve well, my son Methuselah, the books written by thy father; that thou mayest transmit them to future generations. Wisdom have I given to thee, to thy children, and thy posterity, that they may transmit to their children, for generations for ever, this wisdom in their thoughts. Those who comprehend it may not slumber, but hear with their ears; that they may learn this wisdom. The joy taken from this wisdom will surpass the joy deemed worthy of eating some delicious food! (Book 4, Chapter 81)

There is no room for explaining the message given to our current days. Let us just repeat: "I have given wisdom to you, to your children, and even to the children of your children so that they can carry over this wisdom to their own children, to the next generations. The thoughts of this wisdom is beyond their own thoughts. Those who understand will not sleep; they will listen carefully with their ears so that they can learn. The joy taken from this wisdom will surpass the joy of eating some delicious food!"

Enoch's Prophecies on the Past and the Current Times; The Vision of the Weeks

We wanted to include this section about Enoch into this final part, because it was the most surprising section. Therefore, let us analyze the Vision of the Weeks in Book 4, Chapter 92 one by one:

After this, Enoch began to speak from a book. And Enoch said: "Concerning the children of righteousness, concerning the elect of the world, and concerning the plant of righteousness and integrity, concerning these things will I speak, and these things will I explain to you, my children: I who am Enoch. In consequence of that which has been shown to me, from my heavenly vision and from the voice of the holy angels have I acquired knowledge; and from the tablet of heaven have I acquired understanding."

Enoch, then, began to speak from a book, and said: "I have been born the seventh in the first week, while judgment and righteousness prevailed. But after me, in the second week, great wickedness shall arise, and fraud shall spring forth. In that week, the end of the first shall take place, in which mankind shall be safe.

We suggest that the "end of the first" in mention is the end of the ice age, which is known as the Deluge, and which occurred around 11 thousand BCE.

But when the first is completed, iniquity shall grow up; and he shall execute the decree upon sinners.

Many different comments can be made about this section. However, we think that what is being told here has to do with the Sumer Civilization beginning around 3800 BCE, and writing being presented to the mankind for the first time.

Afterwards, during the completion of the third week, a man shall be selected as the Plant of Righteous Judgment; and after him, his generations shall be the Plant of Righteousness forever.

Several researchers share the common belief that the man in mention is Abraham. We agree with the theses which claim that Abraham was born on 2123 BCE. The texts refers to the two nations (Arabs from Ismael, and Jews from Isaac) stemming from Abraham's race as the Tree of Justice.

Subsequently, during the completion of the fourth week, the visions of the holy and the righteous shall be seen. An order for all generations shall be prepared, and a habitation shall be made for them. Then, during the completion of the fifth week, the House of Dominion shall be erected forever.

We can easily follow the resemblance of the sentence "An order for all generations shall be prepared." with the Ten Commandments of Moses, which we think corresponds to 1400s BCE. And "The House of Dominion" is a reference to Solomon's Temple, which was built in 957 BCE. When we look at the last three weeks, we see that there are periods of five hundred years in between each. We think that the following weeks to come will follow the same trait, and have a period of five hundred years in between.

After that, in the sixth week, all those who are in it shall be blind, the hearts of all of them shall be forgetful of wisdom, and in it shall a man ascend. And during the process, he shall burn the House of Dominion with fire, and all the race of the elect root shall be dispersed.

We had guessed more or less that the sixth week would be around 500 - 600 BCE. Among the important historic events of the period are; the Babylonian King Nebuchadnezzar occupying

Jerusalem, burning Solomon's Temple, and sending all Jews on exile, just as told by Enoch. Therefore, let us see what Enoch says about the period at the beginning of the Gregorian calendar.

Afterwards, in the seventh week, a perverse generation shall arise; abundant shall be its deeds, and all its deeds perverse. Then, at the completion of this period, the righteous shall be selected from the everlasting Plant of Righteousness; and to them shall be given the sevenfold doctrine of his whole creation.

Enoch says that selected righteous ones will be determined so that they can transfer a sevenfold doctrine during the seventh week. There is no need to dig in detail into what happened on the zero point of history, which we call as the Common Era, because, as everyone can follow, our subject is the doctrine of Jesus and his apostles, which tells us about the seven layers of heavens and the underworld.

Once everything fitted well into its place up until this point, we wondered in curiosity and excitement what would happen in the next five hundred year periods. However, once we saw the interesting note from the translator, we were disappointed. The note said exactly this:

"At this point, the weeks continue in different versions. We did not include these sections in here."

First, we thought it was a joke! Then, we tried to figure out what was going on. Two pages of a thousands of years old book were not cited, because they had "continued in different versions"! They were perhaps the most important pages, because we believed that this section included prophecies for periods of five hundred years. When we found and translated the excluded parts from the originals, we reached at these sentences:

Afterwards there shall be another week, the eighth of righteousness, to which shall be given a sword to execute judgment and justice upon all oppressors.

We leave this part about the oppressors being judged with the sword of righteousness, which coincides to the period between 500 and 600 ACE, to the discretion of the readers, and move on to 1000s ACE:

Sinners shall be delivered up into the hands of the righteous, who during its completion shall acquire habitations by their righteousness; and the house of the great king shall be open for celebrations. After this, in the ninth week, shall the judgment of righteousness be revealed to the whole world.

When we looked at the famous people of the period one by one, to find out who the person, who had signed as the "Great King", might have been, eventually, we concentrated on one name, who had changed the course of the world: Genghis Khan. So, what did the prophecy written for 1500s say?

Every work of unethical conduct shall appear in the whole earth; the world shall be marked for destruction; and all men shall be on the lookout for the path of integrity.

While analyzing this section, let us remember a sentence from the section on Admiral Byrd: Enki was saying "The dark ages have ended in the last five hundred years." Additionally, looking at 1500s ACE, we guess that the mention is not about one person, but rather a chain of events. While saying "all men shall be on the lookout for the path of integrity", could it be referring to the period of change which began with the Renaissance and Reform movements? Who knows? Right, so what to expect in the current times? Let us see what Enoch's prophecy was for the 2000s ACE:

> *And after this, on the seventh day of the tenth week, there shall be an everlasting judgment. And a spacious eternal heaven shall spring forth in the midst of the angels. The former heaven shall depart and pass away. A new heaven shall appear. All the celestial powers will shine with sevenfold splendor forever. Afterwards likewise shall there be many weeks, which shall externally exist in goodness and in righteousness. Neither shall sin be named there forever and forever.*

Finally, we have reached to the 2000s; the current times. We can only make guesses on what "former heaven" refers to. Former heaven may be referring to Jerusalem, and what is to be extinguished might be the Al-Aqsa Mosque, because, when we look into the past, and to the course of current events, one can say that all the Jews have been preparing for a long time for Solomon's Temple to be rebuilt.

The first temple was built upon the orders of Solomon in 957 BCE, and it was destroyed by the Babylonian King Nebuchadnezzar in 587 BCE. The Jews on exile were aching for the rebuilding of the temple. The Jews, who got hope following the Persian King Cyrus' destruction of Babylon in 539 BCE, forwarded this yearning to Cyrus, and got a positive response from him. The Jews could come to Jerusalem, and build their second temple in 516 BCE following the issue of the proclamation known as the Cyrus Cylinder, which is on display in the British Museum. (8)

The second temple of the Jews has been destroyed by the Romans in 70 ACE. The Jews, who are wishing for building the third temple, have established the State of Israel on 1948 ACE, and following the Six-day War in 1966, they have fortified their presence in the region. Since then, they have been hollowing the ground under the Al-Aqsa Mosque, preparing for the rebuilding of the third temple. By looking at the political balances in the region, we can see these efforts.

The period in front of us cannot be known, but according to Sheik Raed Salah, the Palestinian guard of the Al-Aqsa Mosque, Israel has laid bombs under in order to demolish it in 2020. Based on the news published in the Sabah Newspaper on May 21, 2010, Sheik Raed Salah continues like this: "Israel has been carrying out excavation work under the Al-Aqsa Mosque since 1967. The Haram al-Sharif region, where the Al-Aqsa Mosque is situated, is under the protection of UNESCO. However, despite all the protection, Israel have emptied the soil under the Al-Aqsa Mosque, created numerous tunnels, and laid bombs in these tunnels. In order to be able to place even more bombs and dynamite, they dug out ditches in these tunnels." (9)

Hakan Albayrak wrote in his column in the Karar Newspaper on July 20, 2017 that Israel has been digging and creating tunnels under the Al-Aqsa Mosque for many years in the name of "archaeological excavations", and he added: "One day, we may wake up to the news that the Al-Aqsa Mosque has collapsed." (10)

No one knows whether the destruction will happen in 2020, or at a time closer or further, but there is one thing we know: it will not be a surprise to see the Temple of Solomon being built in lieu, following the demolition of the Al-Aqsa Mosque.

We have attempted to interpret Enoch's Book in the light of our knowledge. Based on the result we have gathered from the prophecies section, great changes will be appearing within this century. The book almost gives the good news that a period of peace will be coming following all the negativities to be experienced. However, it is impossible to say something with preciseness about the accuracy or falsehood of the information in it.

Sources:

1. http://www.hermeskitap.com/catalog/product_info.php?manufacturers_id=10&products_id=58854
2. http://www.peru-explorer.com/hiram_bingham_machu_picchu.htm
3. http://www.cumhuriyet.com.tr/haber/uzay/797112/NASA_dan_carpici_aciklama__Dunyanin_manyetik_alaninda_gizli_gecitler_var.html
4. Gılgamış Destanı 11. Tablet, http://www.piney.com/Gil11.html
5. Yaratılış Kitabı 9-12/16, http://www.yolgosterici.com/tevrat/tevrat01.htm
6. http://www.cbs.com/shows/person_of_interest
7. http://gokturkramu.blogspot.com.tr/2015/05/amon-rann-oglu-buyuk-iskender-mo-356-mo.html
8. https://en.wikipedia.org/wiki/Cyrus_Cylinder
9. http://www.sabah.com.tr/dunya/2010/05/22/israil_2020de_mescidi_aksayi_yikacak
10. http://www.karar.com/yazarlar/hakan-albayrak/bu-gidisle-mescid-i-aksanin-yikilisini-da-normal-karsilayacagiz-4503

SECTION 7

SHAMS, MEVLANA AND THE ANUNNAKI

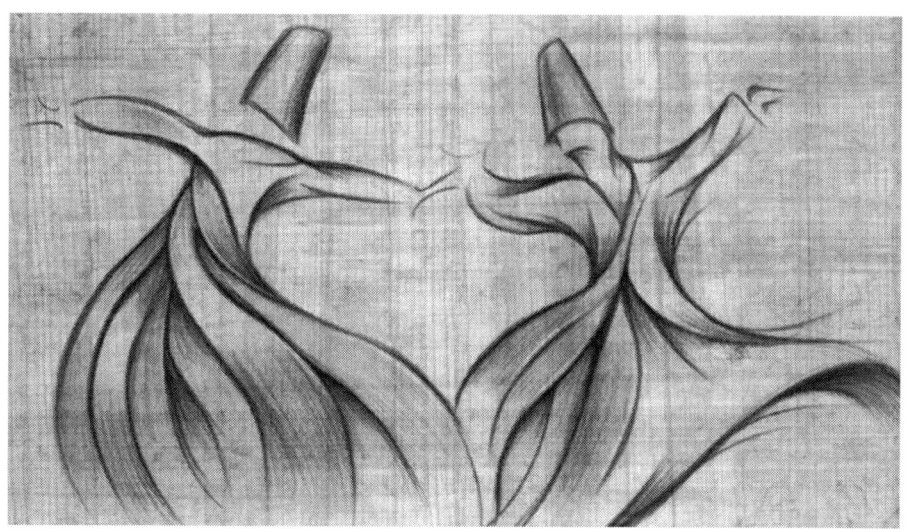

*Oh you, who has a throne beyond existence
and absence!
You are leaving the ground of existence,
don't do!*

From Mevlana to Shams

The Secret of Shams

Who Enoch or Hanok was, and why he did not die, and was lifted to the heavens are still the subjects of discussions. It is thought that Enoch is referred to as Enki-Me, that is; "The One who Enki loves" in Sumer, and as Idris among the Muslims. Sometimes, he is thought to be Thoth and Hermes, but we think he has nothing to do with them, because Enoch has been accepted among the Anunnaki when he was human whereas Thoth is an Anunnaki. While the discussion on whether Enoch is a human being, or Anunnaki, or half-Anunnaki continues, we see in our own culture mysterious characters such as Kithr (Hizir), Zulqarneyn and Shams. Out of these personalities each of whom is a subject for a book, Shams especially attracts attention. We cannot speak with clarity of the physical existence of the other two on Earth, but Shams has lived among people in person. In this part of our book, we will be questioning who Shams; one of the special people, actually is. He is sometimes referred to as the "Flying Shams ad-Din" based upon the notion that "he could have traveled to all those places only by flying".

Due to the mystery around it, the relationship between Mevlana and Shams is one of the most discussed subjects even today. Mevlana's love for Shams sits in the focus of these discussions. The fact that we know so little about Shams increases the mysteriousness.

Mevlana, who was considered to be the most important intellectual of his times in the region, suddenly adopted an admiration for someone he has just met, and alienated himself from the outer world through exaggeration of this admiration. This is not a comprehendible situation for neither the people of his times nor the people of our times. There are too many interpretations and claims on this matter. One of the prevailing views is that Shams is a preterhuman being, and that Mevlana became aware of this.

Sultan Walad, Mevlana's older son, talks of Shams as a preterhuman at the service of God just like Kithr when he mentions "And his Kithr was Shams of Tabriz." (1) And he likens the encounter of Mevlana and Shams to that of Moses and Kithr. According to Walad, Mevlana represents Moses, and Shams represents Kithr:

"All of a sudden, Shams ad-Din came, reached at him. Mevlana's shadow disappeared in his light... Shams said: 'You are advanced in the world of spiritualism, but I am more spiritual than the most spiritual. I am the secret of secrets, the heavenly light of the heavenly lights. The saints cannot reach at my secrets... Love alive becomes dead at worship...' Shams invited Mevlana into a surprising world; such a world that neither the Turk nor the Arab has seen... The Master Sheik came into a state of acquiring new knowledge; he started studying in his presence everyday. He had reached the end, and the study began all over again. While he was the one being abided (by others), he became the one abiding (to him). He was the one and only man as the master of knowledge, but the knowledge he showed to him was novel. A real lover can rarely be found; s/he is hidden from the people like a secret; just like pearls, s/he is hardly found on Earth; just a few people can see his/her indications, and receive news from him/her." (2)

Osman Nuri Kucuk, the Director of the Mevlana Research Center at the Selchuk University, explains the status of "Kithrism" like this: "Certain characteristics of Shams such as; usually acting in contrast to the norms of the society of his times; not socializing with many people; and generally staying alone, bring to our minds Kithr, who does not see himself responsible for what he does against the people and their customs, because the behaviors of Kithr, which are being directed from the vertical dimension, are contrary to the societal customs and norms of the horizontal dimension; they are beyond the conventional molds of knowledge." (3)

Birol Bicer, questions the identification, character and personality of Shams so well in his article with a title "The Sufi with an Anarchist Spirit; Shams Tabrizi", and he displays his difference from the other sufis strikingly with the "Anarchist Spirit". (4)

Prof. Dr. Hulya Kucuk, emphasizes the most profound characteristic of Shams, which was not understood or observed by anyone but Mevlana, as such: "According to Mevlana, Shams is honored with the secret of Vahdet-i Mutlak (the state of being one with God in its absolute sense)."

Dr. Semih Ceyhan describes him like this: "According to Sipehsalar, 'Shams is a dervish, who does not have an inclination for miracles; who wears black felt, and hides himself from the public; who is always at a status of struggles; who travels continuously dressed up like a merchant; who overnights in caravanserais rather than madrasas and dervish lodges; who locks the door of his cell with a strong lock and key although there is nothing in it; who is full of secrets; and who makes a living by weaving waistbands for shintiyans.'"

According to Asst. Prof. Mustafa Cakmaklıoglu: "Shams is described as a personality, who has an ecstatic soul; who can arouse a spiritual impact on the others; who shocks people with his comments and words; who is not understood easily; who refrains from being recognized; who never attaches himself to the sheiks despite meeting with them all the time; who changes his location very frequently; and who is a sufi not afraid of being criticized.

He displays the portrait of sufi, who is never afraid of being criticized by the others upon his fervent behaviors; who has adopted the view that divine love is the secret of spiritual perfection; who thinks that the Perfect Person is the ultimate goal; who is far from forms and ceremonies; who continuously advises to go for quest, and refrain from imitating. In short, Shams is a personality who is alienated even in his own town, and to whom everyone else, including even his own father, is a stranger."

Now, in the light of these explanations, let us think over the relationship of Mevlana and Shams, and the characteristics of Shams that leave Mevlana in awe:

- Mevlana goes into a state of learning new knowledge every single day.

- He begins listening to lectures every day in the presence of Shams.

- Mevlana had already become an intellectual, who had reached the end. Yet, he begins studying once again.

- While he was the one being abided (by others), Mevlana becomes the one abiding to Shams.

- Although Mevlana was a man matured in knowledge, the all novel knowledge Shams shows to him shakes him profoundly.

- According to Mevlana, Shams is honored with the secret of Vahdet-i Mutlak (the state of being one with God in its absolute sense).

- Mevlana sees himself as the "privy to the heavenly light of the Lord", and Shams as the "privy to the secrets of the Lord."

- When Mevlana meets Shams "The secrets became revealed to him like daylight.", and through this meeting, "He had seen what was impossible to be seen, and heard what was impossible to be heard."

It would not be an exaggeration to think that Shams is a preterhuman being once all of these are put together. The researchers, who think in this line, place Shams in the same special ranks as Kithr, and conclude that Shams influenced Mevlana in this way. However, we have a quite different approach to this matter. Those of our readers, who have followed our books and articles, know that we seek a technology "that was not known to us in those times" in all the strange and mysterious incidents. If it can all be explained through science and technology, then, there is no secret, but only an intervention by the Anunnaki. Because we try to bring out through evidence in every platform that the Anunnaki are not at a divine level, but rather they are deified simply due to the technology and advanced science they display, we shall be approaching the relation between Mevlana and Shams with the same point of view.

The Meeting of Mevlana and Shams

On a day when Shams was seated by the door, Mevlana passes by with his students. Mevlana glances at Shams, who seems to be a stranger as conveyed through his dress code; greets him, and keeps going on his route. At that moment, Shams gets up on his feet and grasps the bridle of the horse Mevlana was dismounted on. Mevlana stops the horse, and asks, turning towards Shams: "I beg your pardon. Would you like to say something?" When Shams says, "I would like to learn your name." he replies "Jelaluddin Mohammed". Then, Shams asks, "Who was greater, Bayezid-i Bistami or Mohammad the Prophet?" Coming across with such a question for the first time, he replies: "Of course, the Prophet Mohammad was greater. All creatures and Bayezid have been created in his glory." Expecting this answer Shams asked: "So, how do you explain then: Did not the Prophet Mohammed say 'Oh Lord, we have not understood Thee as Thou deservest to be understood!' whereas Bayezid-i Bistami exclaimed 'Subhani, how great is my glory!'"

Mevlana answers: "Our Dear Prophet's holy heart was such an ocean that no matter how much merit, divine love, affection, and love of the Great God fills into it, it will embrace and surround it. In fact, he even asked for more: 'Oh Lord, increase this blessing you have granted even more!' But, Bayezid-i Bistami's heart was not so wide-open; it would fill up and overflow with the smallest amount of divine enlightenment.

And when it was overflown with a little enlightenment, he would say such things."

Struck with admiration upon this explanation, Shams cried to God and fell to the ground, fainting. Mevlana dismounts from his horse, drops down to his knees, embraces Shams and pulls him up on his feet. In this way, Mevlana and Shams get to know each other. At this time, Prophet Mevlana was 38 years old, and Shams was 60. (5)

Now, looking at this meeting, we cannot see a behavior of Shams which would make Mevlana admire him. Yet, during that period, everyone was admiring Mevlana. Therefore, someone fainting upon his answer can be considered quite standard. At this point, there could be two alternatives: Either Mevlana had been prepared through a vision, dream, etc... prior to this meeting, and he remembered Shans right away when he embraced him; or, Mevlana was surprised by a detail, piece of information, equipment, etc. which was only visible to him. All sources mention that Mevlana went through a profound change after this meeting. Sezai Karakoc says this about this "change": "Some people think that a great change after a certain period suddenly occurs in Mevlana as if a completely different Mevlana was born. However, no change or occurrence happens all of a sudden. Deep down, as part of the invisible plan, a gradual and long period of preparation goes on. Just like an earthquake! We think that an earthquake happens in a flash."

Those who think in the same line as Sezai Karakoc, claim that Mevlana had undergone a period of preparation, and he had known that Shams would be arriving. In fact, such ideas support the view that Shams is not human. If that is not the case, and he notices Shams on the spur of the moment, this leads us to the same result anyway. This means that there is an external intervention not visible to the humans. Well, then; if Shams is not human, what is he? Is he a "hizir" (deus ex machina) or an angel, or an Anunnaki? If we assume that the shortest Anunnaki is about two meters in height, the fact that Shams was a big, tall man may give us a hint. (6)

At the first encounter, Mevlana did not understand that Shams was a person with a preterhuman quality. However, later on, he became clearly aware of this. A similar incident had happened to Prophet Jacob. Although Jacob was alone when he crossed the shallow part of the river Jabbok, he had met with "a man" later on, and he had wrestled with him till daybreak. However, when he had understood that he could not defeat Jacob, he would touch the socket of Jacob's hip, and cause his hip to be wrenched. Then, when "the Man" said "Let me go, for it is daybreak.", Jacob, who understood that this man was a preterhuman (possibly an Anunnaki), would tell him: "I will not let you go unless you bless me." This is told in the Book of Genesis 32:22-32. Now, let us have a look at the further conversations: (7)

The man asked him, "What is your name?"

"Jacob," he answered.

Then the man said, "Your name will no longer be Jacob, but Israel, because you have struggled with God and with humans and have overcome."

Jacob said, "Please tell me your name."

But he replied, "Why do you ask my name?" Then he blessed him there.

So Jacob called the place Peniel, saying, "It is because I saw God face to face, and yet my life was spared."

Most probably, Prophet Jacob had met with an Anunnaki, and just like Mevlana, he had not become aware of it. From that day on, Jacob had been called "Israel", which means "the one who wrestles with God". And, the area where he wrestled took the name "Peniel" which means "the face of God".

The mysterious relationship between Shams and Mevlana

After their introduction, Mevlana takes Shams to the house of Salahaddin Zerkub, one of his distinguished disciples. They cloister themselves in a room for forty days according to some, or for three months according to others. Mevlana says this interesting words to Shams: "Oh, my venerable master! Our home is not worthy of you, but I will try to be a loyal slave for Your Honor. Whatever the slave has belongs to his Master. From now on, this house is yours, and my children are your children." Right at this point, once again, we make a connection with the past:

Book of Genesis, Chapter 19 talks about two angels (Anunnaki) coming into Lot's house. When he sees them, he gets up to meet them and bows down with his face to the ground, and using the same manners as Mevlana, he speaks to them: *"My lords, please turn aside to your servant's house. You can wash your feet and spend the night and then go on your way early in the morning."*

Here, Lot recognizes the angels right away unlike Jacob and Mevlana. It must have been the way they were dressed or the equipment they carried on them. As the incident unfolds, if we put aside the tallness and long lifespans, a development proves that the Anunnaki are no different than humans: Lot prepares a meal for them, and bakes a no-yeast bread. Then, they eat it all together.

As far as we learn from the Torah, young and old alike, all men from all the quarters of Sodom come together and surround the house. They call on to Lot: *"Where are the men who came to you tonight? Bring them out to us so that we can have sex with them."* At this point, the wording as "the men", once again proves that the Anunnaki are in the form of humans.

But Lot begs to them: *"Look, I have two daughters who have never slept with a man. Let me bring them out to you, and you can do what you like with them. But don't do anything to these men, for they are my guests. They have come under the protection of my roof."*

But the men do not comply. They keep bringing pressure on Lot and move forward to break down the door. Right at that moment, the men inside pull Lot inside next to themselves, and close the door. The angels turn every man at the door —young and old— blind such that they cannot find the door.

This incident told innocently by the Torah writers by using homosexuality is actually witnessed in inscriptions such as Kedorlaumer and the Epic Poem of Erra. According to ancient sources, in fact, a war of gods is being experienced there, and cities like Sodom and Gomorrah are fierce Marduk supporters. Erra and Ishum (Nergal and Ninurta) have actually destroyed these five cities because of this. And those who saved Lot were Abraham, and those who had chosen Lot as a soldier.

I don't know if the sentence "The angels turn every man at the door —young and old— blind." is reminiscent of a smoke bomb or beam weapon for you, too. Maybe they used a different technological equipment. Whatever the equipment used was, whoever saw it ran away. The two beings appearing as two angels to whomever saw them points out to the flying capabilities of these "angels" over the Earth's skies.

It is interesting that Shams also has a similar incident, and this has reached to our times through the written records of those who have been witness to it. One day Mevlana and Shams are listening to the lyre in an assembly. Someone says, "The sound of dervishes and lyre…" Shams turns to him, and replies: "You do not see nor hear." Consequently, the man grasps his throat with his hand, and then becomes blind and deaf. (9)

Another similar incident is told: Shams wants to stop by and sleep at an inn on his way from Damascus to Konya, but when he finds out that every inn he stops by is fully occupied, he thinks of overnighting at a mosque. He goes to the mosque, and prays with the public for the night prayers. He takes off his cloak, places it under his head, and lies down. After a while, the man on duty, whose job was to lock the doors of the mosque, comes by, and upon seeing that someone was lying down he says "You cannot lie down here. Get up!" Shams replies sitting up: "I will cause no harm to anyone. I'm a stranger here; I come from a long way. The inns have no availability, and I have no other place to stay. Let me stay here till the morning." When the man who was there to lock the mosque replies to him "Do not cause me trouble. I told you to get out of here. Otherwise, I know how to kick you out." Shams takes his cloak and leaves in silence. The man on duty begins to fight for his breath while looking behind him, and screams "Help! I'm suffocating!" Upon hearing his screams, the imam comes by, and asks "What is going on? Why are you screaming?" When the man tells about the situation, the imam leaves the mosque right away, and runs after Shams. When he catches up with him, he says "Sir, he is an ignorant man; he has no manners. Please, forgive him!" Shams looks at the imam, and replies: "His business is out of my hands. There is nothing I can do. I can only pray that he dies in faith." (10)

As seen in these two incidents, the experiences mentioned are indicators of someone with a preterhuman ability rather than a saint, who can utilize this ability easily even though it may be whimsical. One of Mevlana's stories displays Shams' tough and offending personality anyway:

The public, who does not understand Shams, gossips about him, because his words are stinging and offending. He tells the truth directly. One day, Mevlana strolls around the marketplace together with his dervishes. He comes across with a dog feeding her puppies, and says: "Look, my dear dervishes! How much they resemble to us!" Before the dervishes could say more than "God forbid! How could it be, Sir..." Mevlana says: "Watch this mother breaking down the bone into dust to feed her puppies. Had it not been for the mother, the poor puppies would try to bite the bone, break their teeth, and die of hunger.

Similarly, do not take Shams' tough words into account. Take into account how I turn those words into milk for you."

The envied bonding between Mevlana and Shams

Mevlana never leaves Shams starting with the very first day he meets him; he no longer goes to give his lectures for his students, the preaching at the mosques, and to attend the conversations. The only one, who comes next to them during the special conversations Mevlana and Shams were having, is Mevlana's older son Sultan Walad, and that is to serve to them. We are wondering what made the conversation inside so attractive; what it was that Shams was telling, which made Mevlana so affected, or what he made Mevlana experience.

Although the sources tell that every day they go into contemplation on the creations of God, perform the ritual prayers of Islam, and refresh their conversations by mentioning the name of God, these are far from being satisfactory. After all, Mevlana could have done these together with so many other people. At this point, with the hope of finding different hints, we should be digging further into the recounts of Mevlana's family members, who were the greatest witnesses to this interesting friendship, which has made history. Sometimes the strange sounds heard behind closed doors, and at other times, incidents witnessed through the doorways may lead us to a completely different interpretation that is different from the most common ones.

The mysterious visitors of Mevlana and Shams

Mevlana's wife Kira Khatun was an eyewitness to an interesting incident: "One day, in the middle of winter, when Mevlana was sitting together with Shams in seclusion, Kira Khatun wonders what they are doing inside, and tries to listen to the other side of the closed door. All of a sudden, she sees that the wall of the house cracks open, and that six 'imposing' men, who belonged to the invisible world, enters inside, greets, kisses the ground, and places a bunch of roses in front of Mevlana. These six imposing men sit there until about mid-day prayer time in complete peace, but they do not talk at all. After the mid-day ritual prayers are performed, those six almighty persons get up in all respect and honoring, and leave through the wall just as they had arrived.

Kira Khatun passes out in the heat of the incident. When she wakes up, Mevlana comes out, and gives the bunch of roses to her to preserve. Kira Khatun picks up a few rose petals, and sends them to a few different herbalists. Since she had never seen such roses before, she was curious to learn what type of roses they were, where they came from, and how they were named. All the herbalists become puzzled by the freshness and scent of the rose. They can't help asking "Where the hell this strange rose appear from in the middle of winter?"

Later on, Kira Khatun's servant takes the rose petals back and returns to home. Then, Mevlana comes in, and says: 'Preserve that bunch of roses well. Do not show it to anybody unrelated, because the gardeners of the sacred Irem Gardens, which are the poles of India, have sent them as a gift to enforce your eyes and heart's mind. Mind you; save it well to ward off the evil eyes on it."

Researchers have various interpretations on this incident. Some say that these six persons are actually angels, and that the roses have a symbolic meaning. Yet, others claim that the area called as the Irem Gardens corresponds to the mysterious underground country of Agartha. (11)

As the witness to the incident is Mevlana's wife it got our attention. The first information Kira Khatun gives is that the wall of the house opens up; and that some people enter inside, and put some roses on the ground. To see into the Anunnaki dimension of the incident, we can check into similar accounts in the ancient information.

A very interesting incident is told in the Bible, Book of Daniel: Chapter 5: Belshazzar, the Babylonian King gives a great feast to a thousand of his noble men. As the feast continues in joy, in front of the eyes of everyone, there appears the fingers of a man's hand, and it begins writing Mene, Mene, Tekel, and Uparshin on the plaster of the wall of the king's palace next to the candlestick. When the king sees the hand, his thoughts trouble him, his countenance changes, he blanches, the joints of his loins are loosened, and his knees smite one against another. The same night, Daniel is called in, and he is asked to interpret what is written. Daniel interprets these words as "God hath numbered the days of thy kingdom, and finished it. Thou art weighed in the balances, and art found inadequate. Thy kingdom is divided into two, and given to the Medes and Persians." In fact, Belshazzar gets slayed on that same night, and the Persian King Kyros takes the country.

A hand without a body that appears in front of the eyes of everyone! Considering that the Anunnaki have figured out the system of frequency and vibrancy, what is being done here is to reflect only the frequency of the hand into the humans' area of visibility by using an equipment or another way. We see the evidence for this technology being used not only for the hand, but also for the whole body in the Bible, Book 1 Samuel: Chapter 3:

1 Samuel 3:2: One night Eli, whose eyes were becoming so weak that he could barely see, was lying down in his usual place.

1 Samuel 3:3: The lamp of God had not yet gone out, and Samuel was lying down in the Temple of the Lord, where the Chest of God was.

1 Samuel 3:4: Then the Lord called Samuel. Samuel answered, "Here I am."

1 Samuel 3:5: *And he ran to Eli and said, "Here I am; you called me." But Eli said, "I did not call; go back and lie down." So, he went and lay down.*

1 Samuel 3:6: *Again, the Lord called, "Samuel!" And Samuel got up and went to Eli and said, "Here I am; you called me." "My son," Eli said, "I did not call; go back and lie down."*

1 Samuel 3:7: *Now Samuel did not yet know the Lord: The word of the Lord had not yet been revealed to him.*

1 Samuel 3:8: *A third time the Lord called, "Samuel!" And Samuel got up and went to Eli and said, "Here I am; you called me." Then Eli realized that the Lord was calling the boy.*

1 Samuel 3:9: *So, Eli told Samuel, "Go and lie down, and if he calls you again, say, 'Speak, Lord, for your servant is listening.'" So, Samuel went and lay down in his place.*

1 Samuel 3:10: *The Lord came and stood there, calling as at the other times, "Samuel! Samuel!" Then Samuel said, "Speak, for your servant is listening."*

Samuel 3:11: *And the Lord said to Samuel: "See, I am about to do something in Israel that will make the ears of everyone who hears about it tingle.*

    Here, Samuel hears supernatural voices, and thinks that Eli, who is sleeping in the other room, is calling him. After the incident repeating itself three times, he takes the advice from Eli, and does as he tells him: He replies to the supernatural voice, and starts speaking with someone he does not see. This incident hereby constitutes an example for an Anunnaki in the high frequency communicating with a human being in low frequency.

Sure enough, for people who have no knowledge of the existence of technology, these incidents might seem like miracles. However, to us, who know about the current technology, these remind scenes from sci-fi movies. The realities will be revealed soon when the humans figure out the secrets of vibrancy and frequency.

If we assess the incident Kira Khatun tells from our point of view, we can conclude that the Anunnaki, who have the technology of vibrancy and frequency, has come to visit Shams, who is another Anunnaki just like themselves. The purpose of all these experiences including the extraordinary gift of roses is to impress Mevlana. Now, let us look into another incident:

The sand of Hejaz on Mevlana's feet

Mevlana's wife Kira Khatun tells that one night Mevlana disappears. Although Kira Khatun looks for him everywhere in the madrasah, she cannot find a trace of him. And she sees that all the gates are closed anyway. Kira Khatun tells the rest of the incident like this: We were all caught in surprise with this. After everyone had slept, I woke up suddenly. I saw Mevlana performing the ritual prayers. I did not say a word till he finished. Once he finished with his prayers, I noticed that his feet were covered in dust. I also saw that there were colored sand grains in between his toes. I asked him about this situation in complete fear. He gave this answer to me: "There was a saintly dervish at the Qabe, who continuously talked about our love. I went to visit him for a while. And this is the sand of Hejaz; hide it; do not mention it to anyone." (12)

What initially comes to mind for the technology used here is the short-range spacecrafts mentioned in the ancient tablets as celestial boats, which we call as UFOs nowadays. Enoch's Book mentions that Enoch was taken to several parts of the world on celestial chariots. Who knows; maybe Shams and Mevlana took a journey in a similar way...

Once we put all of these together, we cannot agree with the idea that Shams and Mevlana simply performed the prayer rituals and conversated. According to us, what happened is that Shams, who had superior information, gave Mevlana such information, showed him such visions or made him go through such experiences that he was in complete admiration of him. If Shams had showed Mevlana one of the tablets, which is a fundamental toy for our kids nowadays, and had him watch a video on it, this would have been a huge miracle for Mevlana. For an intellectual such as Mevlana, who gives utmost importance to research and knowledge, to say the following words after Shams' departure, is quite crucial in this context:

"My heart/mind found the glitter of life; it opened up and cracked. My heart/mind found your atlas; I have turned into an enemy of that rag bag. I have been busy with books and notebooks just like Mercury. I had sat on the top ranks of all the literary men. But, when I saw the forehead plate of the cupbearer, I lost myself; I broke all the pencils in my hand."

The departure of Shams

By then, Mevlana leaves aside all his preaching, lectures, duties, and responsibilities; in short, he stops doing every behavior and action. He casts outs all the books he had been reading daily. He stops calling his friends and followers. Everywhere around Konya, a wind of protest and rebellion against this new situation blows. When those people who had never welcomed the existence of Shams started talking smack about Mevlana, this is how Mevlana replied to them:

"Before his light shined on me, I was a dead piece of embroidery on your stone walls. Before he bowed on my strings, I was a dry lyre, which was a stranger to itself, and which played and sang the same tune all the time. I see in his palm, vineyards and gardens; and waters as vast as the oceans, and as clear as the

oceans. I rest under the shadows of the trees growing in his palm. However, you cannot see any of these."

If we consider Mevlana's words in the light of the current technology, we can more or less guess the technology Shams uses behind closed doors.

When things get into such a grievous condition, one day, Shams reads a verse from the Qur'an to Mevlana. The verse interprets as "and this is the separation between you and me". This separation becomes true; one night, Shams leaves Konya without notice, and goes to Damascus. Shams' departure makes Mevlana extremely unhappy. From then on, he does not want to see anybody, and he does not accept any visitors. He is now off his oats, and he completely stops visiting the samah (whirling) lodges, and meetings with friends. Mevlana recites odes full of longing and love, and tells about Shams-i Tabrizi through messengers he sends everywhere. Some of the followers feel remorse, and ask for forgiveness from Mevlana, but some others get even angrier, and nurse a grudge against him. Months pass by like this, and when Mevlana understands that he cannot go on like this anymore, he decides to send his son Sultan Walad to Damascus. Sultan Walad finishes his preparations, and starts off. He finds Shams talking to a young person at the inn his father had described to him. He explains the situation as much as he can. Shams accepts the invitation; returns to Konya, and meets with Mevlana. However, the gossiping goes on, and one day, Shams says to Sultan Walad "Oh, Walad! Once again, they have prejudices against me. They have unanimously decided to separate me from Mevlana. The pain of separation will be very deep this time." Rumors say that Shams disappears all of a sudden on a day in 1247 ACE. Although some rumors claim that Mevlana's son Aladdin killed Shams, there is no proof to support this.

After Shams disappears, Mevlana goes to Damascus twice, but he cannot find him. Although the dates of these last journeys are not known for sure, most probably they are between 1248 and 1250. As expressed by Sultan Walad; Mevlana had not been able to find Shams physically, but he had found him (Shams) in himself (Mevlana) in terms of meaning. He sees Shams appearing in his own existence just like the moon, and says "I am separated from him bodily; but without bodies and lives, both of us are heavenly lights. Oh, the searching one! If you wish, see him! If you wish, see me! I am Him; and He is me..." (13)

After Shams' departure, Mevlana lives in seclusion for many years. During the following years, although Salahaddin Zerkubi and Husamaddin Chelebi tried filling in for Shams-i Tabrizi, Mevlana, who summarizes his life with three words "I was raw [khâm], I became cooked [pokhta], I was burnt [sokht]." dies on December 17, 1273.

Did Shams die? Or did he get slayed? Or did he leave?

Following the decision of the Higher Education Council in 2012 for the establishment of the Department of Research on Mevlana and Mevlevism at the Seljuck University Institute of Research on Mevlana, "Mevlana and Mevlevism" has become a field of science. Let us have a look at the statements of Associate Prof. Dr. Nuri Simsekler, who is the director of the Institute of Research on Mevlana, which has become the only scientific authority in Turkey after the decision in 2012: (14)

"The only information about this incident of murder comes from Aflaki's "The Feats of the Knowers of God", which was written one hundred years after this date. This work mentions Shams being killed. However, if we take into account other information from the same source, Shams' burial place should not be at the present location, but rather right next to Mevlana's Father's burial place, which is presently in the Mevlana Museum. This is a contradiction. The present day tomb of Shams-i Tabrizi had been built on the location of a water source, which both the Christians and the Muslims living in Konya believed to be healing, one hundred and fifty years after the period Mevlana lived. Not only whether or not Shams was murdered, and thrown into a well here, but also his burial place is not known for sure. There are simply too many rumors about this matter; three different locations in Konya, four all around Turkey, and other locations even in Iran and Pakistan are perceived to be the burial place or the chair of Mevlana.

We consider the most accurate and the earliest source to be the Ibtidaname of Sultan Walad, who is the son of Mevlana. There are only expressions such as 'hid, disappeared'. For the time being, we think this is the truth. We try to pick some things out of the poems of Prophet Mevlana just like a detective. However, in certain verses Mevlana uses expressions such as 'Our Shams is fallen into the well.', and yet in others 'Who says that he is dead?' Science is left without any material against the question "Was Shams-i Tabrizi murdered?" For the time being, with all the data at hand, it is not possible to say 'He was murdered.' or "He was not murdered.' Had Mevlana known that Shams-i Tabrizi was murdered, or had died, he would not have gone twice to Damascus to look for him.

We have reached at the conclusion that Shams did not die, and he was not murdered. He just left...

Makalat

Makalat is the only work by Shams that is known. However, this book was not personally written by Shams. The work not only gives information about the lives and ideologies of Shams and Mevlana, but also information that is vivid and pertaining to its period. There are strong ties between Makalat and Masnawi. Although it is not for sure, it may have been compiled by a group of Mevlana's followers, including his son Sultan Walad in the first ranks. Mevlana has put together many of the stories, matters, and expressions of Makalat in his Masnawi. (15)

If Shams is an Anunnaki, why might he have visited Mevlana?

If Shams is an Anunnaki, why might he have visited Mevlana? Or, looking through another window; why was teaching Mevlana so important for this Anunnaki? It is not possible to give an exact answer. However, we can make several comments considering the popularity of Mevlana's ideas and views even today in Turkey and around the World.

Mevlana, known as "Rumi" in US, is one of the most popular names, even in an environment in which the discussions on Islam have been revived. (16)

Jesse Kornbluth, a commentator for the internet newspaper Huffington Post, has written this for Rumi: "His funeral lasted 40 days, and he was mourned by Christians, Jews, Muslims, Persians and Greeks. Rumi belongs to everyone. And always will. It makes perfect sense that this 13th century Muslim is now said to be the best-selling poet in 21st century America." (17)

Lee Briccetti, the executive director of Poets House, an institution in US, which has initiated a library series about Rumi, says: "Across time, place and culture, Rumi's poems articulate what it feels like to be alive. And they help us understand our own search for love and the ecstatic in the coil of daily life.", and she compares Rumi's work to Shakespeare's for its "resonance and beauty".

Coleman Barks, who has initiated a Rumi Renaissance in US through his translations, and who has made him the best-seller poet, explains the reason behind Rumi's endurance until today through his astonishing freshness in creativity, the deep longing surfacing in his poems, and his humor mingled with wisdom. Barks' translations of Rumi have reached twenty-two volumes in thirty-three years, and the Rumi books published by HarperOne Publishers have been translated into twenty-three languages, and have sold over two million copies. (18)

In these days when we are on the eve of entering the Age of Aquarius, if the whole world is still speaking about the wisdom that existed eight hundred years ago, and even further; if the whole world is recognizing Mevlana —as he is called in our country— as "Rumi", we can say without hesitation that this mission and vision is to guide the humans in the new order.

Let us pose a second question; this time about Shams: If Shams is an Anunnaki, which one is he? If we seek the answer to this question through examining Shams' behaviors and actions, unluckily, we cannot get a result, because he carries all the characteristics of an Anunnaki. However, we may find a hint, if we look at the geographical location he comes from, and the adjective he chooses for himself.

In this case, let us have a look at the city of Tabriz, which he chose to use in his name. There is no absolute information on when Tabriz was founded. Based on the archaeological excavations in the city, it is thought that the city has a five thousand year old history, and that it had been founded right after Sumer and Egypt. This city, which had been home to Turks almost in every period, is a habitat for the Azeri Turks even today. Considering that these lands belong to the eastern clan of Enlil, we catch a little hint, and ask whether Shams could be one of the members of the Enlil clan.

Shams is derived from "Sham". Shams means "glittering, shiny", but the shining in mention is not referring to the natural radiance like that of the sun's. Consequently, in Arabic, it has been the symbol of the shininess other than that of the sun's, in the context of its meaning as candle, torch. G.M. Red Slob has pointed out in a study published in the German Oriental Society Magazine that the terms shem and shamaim ("the skies/heaven") have been derived from the root word Shamash, which means "that which is highward". On the other hand, Samash, Shamash, or Shem-Esh in Akkadian is more obvious: Esh stands for "fire", and combined with Shem, it means "The Fire that Shines".

The meaning of Utu, who is the son of Sin —an Anunnaki from the Enlil clan— is "that who enters while shining", and it corresponds to Shamash in Akkadian. When we say that the Sumerian Sun God Utu/Shamash, who is simultaneously the twin brother of Inanna, is the protective Anunnaki of the Turks, a single question comes to our mind:

We wonder if Shams-i Tabrizi could be our one and only wild Shamash.

Sources:

1. Sultan Veled, 'İbtida-name' XVIII. Bölüm
   http://dosyalar.semazen.net/IBTIDANAME_SULTAN_VELED.pdf
2. Sultan Veled, 'İbtida-name' XCVI. Bölüm
   http://dosyalar.semazen.net/IBTIDANAME_SULTAN_VELED.pdf
3. https://rumimevlevi.com/tr/yazarlar-cemiyeti/141-osman-nuri-kucuk/2640-sems-i-tebrizi-93425217.html
4. Birol Biçer, Anarşist Ruhlu Sufi; Şems-i Tebrizi
   http://www.derki.com/ezoterik/anarsist-ruhlu-sufi-sems-i-tebrizi/
5. https://gizliilimler.tr.gg/%26%23350%3Bems_i-Tebr%EEz%EE.htm
6. http://fahrettin-bayrasa.blogspot.com.tr/2011/10/semsi-tebrizi-hazretleri-kimdir.html
7. http://www.erzurumyenigun.com/semsi-tebrizi-ve-mevlana-1699h.htm
8. http://www.yolgosterici.com/tevrat/tevrat01.htm
9. http://www.yolgosterici.com/tevrat/tevrat01.htm
10. http://muhammedi.net/html/adost_konyavelileri_sems.html
11. http://www.turkiyegazetesi.com.tr/yazarlar/vehbi-tulek/405509.aspx
12. http://biyografiler2013.blogspot.com.tr/2015/06/mevlana-celaleddini-rumi.html
13. http://www.golgelerkitabi.com/forum/index.php?topic=8878.0;wap2
14. http://www.konya.bel.tr/bldfoto/m/mevlana/hayati/bulusma.html
15. http://www.sabah.com.tr/kultur-sanat/2012/11/29/sems-olduruldu-mu
16. http://www.tasavvufdergisi.net/Makaleler/2144280217_14.12.pdf
17. http://www.hurriyet.com.tr/amerika-mevlana-celaleddin-rumiyi-yeniden-kesfetti-15598791
18. http://www.huffingtonpost.com/jesse-kornbluth/the-greatest-muslim-poet_b_687833.html
19. http://www.bbc.com/turkce/ozeldosyalar/2015/01/150119_vert_cul_mevlana

SECTION 8

THE PYRAMID WAR AND THE ILLUMINATI

(8760 BCE)

Adda Cylinder, British Museum – No: 89115

When is the academic community going to accept the fact that the pharaohs of the Fourth Dynasty did not build the great pyramids?

From the movie Stargate 1994

In the last three sections, we have written about the messages given to us through Admiral Byrd, the hidden prophecies about our current times in the Book of Enoch, and how Mevlana's Philosophy forms the fundamentals of the intellectual system of the new age. According to us, all of these are the preparations for the new system to be established by Enki for the Age of Aquarius. From this section on, we will be empasizing the political, cultural and scientific preparations for the new system.

Looking at the current times from a political and economic point of view, we come across with global organizations, giant corporations, and secret sects going beyond borders ruling the World rather than governments. Among these the most well-known is the Illuminati Organization the members of which are the top level executives. How come this organization, which was established in 1776, has come to be ruling the entire world today? Well, we will be seeking the answer to this question in our journey to the past while being fed by the symbols. The ancient information, tablets and inscriptions will lead the way for us on this journey.

Illuminati is a message of the West to the East, who has been ruling the world in the last two thousand and five hundred years. It cannot be understood how the nations in the West of Europe, who have little contribution to the world as a civilization since the times of Sumer, united in the leadership of Britain, and captured the leadership from the East. The western states pioneered by Britain, who had no precedents and preparations known to public, has surpassed the eastern states, who had all the technology; as if touched by a secret and mysterious hand, as never have been witnessed before in the six thousand years of history.

We will be lifting the curtain off on the rise of the West in the next section, but, right now we will focus on the eleven thousand year old foundations of the Illuminati Organization, which plays a role of catalyzer in this rise.

The missing capstone of the Great Pyramid in Egypt appears today as the "eye that sees everything", which is one of the symbols of the Illuminati. So, what is the meaning of this? What kind of a message it gives? And to whom it is intended for? To find the answers for these questions, we will once again travel to the past.

The three pyramids and the Sphinx in Giza: Who do they belong to?

Herodotus writes that he had visited Giza in 5th century BCE, and that the first pyramid was ordered to be built by Khufu (Cheops) based on the information he had gathered from the Egyptian priests. Herodotus attributes the other pyramid to Khafran, and he implies that the smallest one was ordered to be built by Menkaure, aka Mikerinos. The Egyptian priests thought so, because there was the Khufu temple next to the Great Pyramid, the Khafran temple next to the Middle Pyramid, and the Menkaure temple next to the Small Pyramid.

Allegedly, the British Major General Vyse, who put together this information in 1837, wangled a bit; using his incomplete knowledge of hieroglyphs of the times, and the paint which the Arabs call "mogra", he wrote "Khufu" (in fact "Rafu") at the section of the pyramid with an hermetic seal. (1) Since that day, the mainstream science accepts that these three pyramids were ordered to be built by the pharaohs as the kings' tombs.

This action of Major General Vyse, which is claimed to be a fraud, encouraged two British archaeologists in the next century; Charles Dawson and Arthur Smith Woodward. Following a similar path of hoax, these two scientists claimed that they had discovered the missing link between humans and apes. This mid-form, called as the Piltdown Man, found scientific acceptance from 1912 —the year it was discovered— till 1953. Then, the British Natural History Museum accepted that the two scientists had played a trick. (2)

There is no evidence indicating that the three pyramids in Giza have been ordered to be built by Cheops, Khafran, and Mikerinos. We are guessing that these three pyramids and the Sphinx were built right after the Great Flood on 10500 BCE. There are various views on why and how these ancient buildings were created. We, who defend the Ancient Astronauts Theory, believe that they were built as an additional facility for the space ports of the Anunnaki in Sinai, Baalbek, Jerusalem, and Mount Katherine. The spacecrafts were landing depending on digital data. However, if they came across with any problems at landing, they would use the two hills on the right (Mount Katherine hills), and the two artificial hills on the left (Cheops and Khafran) as visual aids, and by placing themselves right in the middle, they could land at the space port in Sinai. Of course, this was not the sole purpose. The pyramids were built with a technology much more superior to that of ours today, and they were serving ten folds of purposes like this.

Right. What about the Sphinx?

It is assumed that the Sphinx was built at the same time with the Second Pyramid during the reign of Pharaoh Khafran. Thus, the view that it was built during the time of the Fourth Dynasty (2723 – 2563 BCE), and that the face of the Sphinx reflects that of Khafran has been adopted. Although this view lacks any basis, it is still accepted in this way by the mainstream science. However, according to the geologist Dr. Robert Schoch from the University of Boston, the age of the Sphinx is much older than what is known.

A meteorological analysis made on the Sphinx and its layers by Dr. Schoch has proved that it was carved out of a piece of rock that existed at the location much before the times of the dynasties of the pharaohs. Dr. Schoch's research methods included these: the seismic screening of the rocks below the surface of the ground, carried out by Dr. Thomas Dobecki, a geophysicist from Houston; the Works of Anthony West, the

Egyptologist from New York; and the analysis of the signs of erosion and level of water on the Sphinx and its surrounding. According to Dr. Schoch, the erosion as a consequence of rain on the pyramid points to a period between 10 thousand and 5 thousand BCE when the climate of Egypt was damper. More interestingly, the next location of visit for Dr. Schoch, who believes that the age of the Sphinx is much older, has been Gobeklitepe.

There is one more evidence indicating that the Sphinx and the three pyramids were built much before the times of the Fourth Dynasty. This piece of evidence named as the Inventory Stela has been discovered in the ruins of the Isis Temple near the Great Pyramid by Auguste Mariette in 1850. According to this stela, which is presently in the Cairo Museum, when Khufu came on to stage, the Great Pyramid had already been standing there for quite a while, and his wife was Goddess Isis. In other words, the Great Pyramid does not belong to Khufu, but to this goddess. Moreover, the Sphinx and the second pyramid, which is attributed to Khafran, had been in their present location already. The rest of the writing describes the position of the Sphinx accurately, and records that it was damaged due to a lightning strike. In short, this stela tells that the Great Pyramid and the Sphinx had already been standing up when Khufu aka Cheops had become the pharaoh.

Everything in the inscription overlaps with the facts known, and as it was later called, the "Inventory Stela" carries all the signs of reality. However, the scientific world, which did not want to rewrite a huge amount of history, announced this stela to be fake. The reason for this was mainly Major General Vyse's alteration of the names of the pyramids about ten years before this; attributing them to Cheops, Khafran and Menkaure.

The pyramids of the Deities and their copies

In reality, these three pyramids have been known to be the pyramids of the deities for many years, and the half-deities had ordered many other pyramids to be built, because they wanted to follow the footsteps of the deities when the Egyptian civilization was founded. Yet, no other pyramid has been as exquisite as the first three pyramids. While many had collapsed, several others proved to be far from the mathematical ratios with their 44 degrees low angles. Still, we think that even these pyramids built by the humans were built with the help of the technology of the Anunnaki. Based on the latest claims, all the pyramids built during the Ancient Egyptian period were built for the purpose of converting energy rather than as burial chambers. Today, when we look into the subject of pyramids, we see that almost all of them are built on a bed of energy such as natural gas, magnesium, gold and water. If this theory is correct, then, we can conclude that the Anunnaki used converted energy derived from the Earth rather than fossil fuels in their spacecrafts. Perhaps, there is a trace of this in the Dendera Temple of Hathor. (Figure 1)

(Figure 1): The mural depiction in the Dendera Hathor Temple

According to us, this mural depiction, which is accepted as the "Hathor Necklace" by the mainstream science, is actually a depiction of how the Earth's reserves are turned into energy. Especially the energy sign used in the product section of the depiction is still being used. (Figure 2)

Figure 2: The mural depiction in the Dendera Hathor Temple

It is thought that the first of the impressive pyramids of Egypt was ordered to be built by Djoser, who was the second pharaoh of the third dynasty, on 2650 BCE. According to mainstream science, Djoser orders Imhotep, a very smart scientist and architect, to build a grand scale burial place. The scientific world turns a blind eye to the question of why Djoser chose to make the building in the shape of a pyramid which had no precedence, but for us, it is quite obvious; Djoser had already three pyramids of the deities in front of his eyes.

Those who visit Saqqara today, can see the remaining parts of this first pyramid built by Imhotep, which is known as the Step Pyramid. (Figure 3) On the other hand, the pyramid trials of the two consecutive pharaohs Sekhemkhet and Khaba were unsuccessful, and their buildings had collapsed.

Figure 3: The Saqqara Pyramid

For some unknown reason, the new pyramid, whose construction had begun during the reign of the next consecutive pharaoh, that is; Huni, and continued by Sneferu, the first king of the fourth dynasty, was intended to be built different than the previous step pyramids. It was tried to be built as a real pyramid with smooth surfaces, and with an outer surface at a 52 degrees angle, which was the angle of the pyramids of the deities. However, this construction also turns into a grand scale fiasco a while after, and collapses. Later on, Sneferu orders the Red Pyramid to be built with the safe angle of 44 degrees.

The scientific world says that the three pyramids in Giza were built right after this pyramid, but this is not satisfactory at all. For example, one of the interesting questions is this: If the Great Pyramid was ordered to be built by Cheops, then, why wasn't the second pyramid built by his son Djedefre, but rather by Khufu who came after him? For those who would like to learn further about this hoax of names, we can suggest Zecharia Sitchin's book "The Stairway to Heaven".

The Great Pyramid War

Zecheria Sitchin claims that "Wars of Gods" were experienced in the Great Pyramid, the biggest of the three pyramids, about 11 thousand years before our times. At the end of this war, the pyramids were all left empty, and the foundations of a war between the east and west clans to last until current times were laid. As far as we can learn from the ancient tablets, this war's sides back then were Ninurta and Marduk, who represented the east and west clans.

The cause of the Pyramid War is recounted as Inanna's fiancé Dumuzi's death as a result of his falling off near the Aswan Dam in Egypt, which we think was an accident. Inanna did not really care whether the death was by accident or not, and she blamed it on Marduk. Getting the support of the east clan, Inanna chases vengeance. Enlil, Ninurta, Adad, Shamash and their subordinate squads were among the supporters of Inanna, but Marduk was not left alone either, by his own clan. Enki and his subordinate squads leading the way, Gibil and Ninagal; two of the sons of Enki, and Isis, and her son Horus were also among the supporters of Marduk. One of the two other sons of Enki; Nergal says he would give support in case of danger, and Thoth, the other of the two, says he would not give any support under any circumstances. Consequently, the fuse of a grand scale war between the two clans was ignited.

This war, which came to be known also as the Wars of Gods, has found wide coverage for itself in the Sumer and Egyptian records as well as other written historic records, and depictions with figures. The hymns attributed to Ninurta make numerous attributes to his success and heroism in this war. The hymn called "Created like Anu" and the majority of the text "The Exploits of Ninurta" tell about this war. (3) However, the main historic record of the war is the inscription called "Lugal-E Ud Melam-Bi" in Samuel Geller's book "Texts and Observations on the East". (4)

The poem, which glorifies Ninurta, his success and his shiny weapon, describes the area where the war is fought as the "mountainous country", and the main enemy as the "great serpent". The mountainous country hereby is the mountainous land in the south of Egypt, and the great serpent is Azag as in the Sumer poem. At another point of the same inscription, this enemy is referred to as Ashar. Both of the epithets belong to Marduk. This inscription shows that the leaders on each side were Ninurta for the east clan, and Marduk for the west clan. The war starts in the mountainous land in the south of Egypt, and continues around the Giza Complex. The second of the thirteen tablets on which this poem is written on talks about Ninurta's flying bird. This interesting bird is most probably a spacecraft.

The leaders of the West Clan, Enki leading the way, retreat and take refuge in the Giza Complex. Marduk continues resistance while being besieged, and upon the events getting so out of hand, he sends his men to call Nergal, his brother in South Africa, causing him to give up on his decision, and be part of the war. J. Bollenrücher's work called "Prayers and Hymns to Nergal" tells about the deities of the Enki clan, which was hemmed in within the Giza Complex.

Nergal breaks through the besiege of Enlil around "E-Kur" (the Great Pyramid) accompanied by his lieutenants, and by entering through "the locked doors which by themselves can open" makes the family of Enki in dire straits to rejoice. He is greeted by a "welcome song" inside. The assistance provided by Nergal were also recorded on a clay cylinder discovered in the ruins of the Enlil Temple in Nippur. In this one, Nergal joins a group defending "the Formidable House Which Is Raised Up Like a Heap" (The Great Pyramid).

In this war, Horus and his army of "Metal People" took their place in the fights, too. We always wonder where and how human beings learned fighting a war. The answer is hidden in this war. Horus learns the art of ironworking from Gibil, and teaches this to human beings, too. He then creates an army to rule over Egypt. Horus gets wounded by a beam bouncing from Ninurta's shiny weapon, and loses one eye in this war. Sumer texts mention Horus as Horon; "the child who does not know his father". Upon this loss of Horus, Ninhursag (Ninmah) plays an intermediary role to end the war.

The ninth tablet of the Lugal-E text begins with a speech in which Ninmah addresses her own son Ninurta: She tells him that she will be going to the "House Where Cord-measuring begins" (the Great Pyramid). But, Ninurta is astounded by Ninmah's decision to "enter alone the *Enemyland*"; but since her mind is made up, he provides her with "clothes that should make her unafraid". Upon getting closer to the area of the Pyramids, Ninmah calls to Enki. Due to the broken parts of the tablets, the conversation between them is unknown, but, the besieged Enki consents to make an agreement in order to stop the war.

Later, Ninmah goes to Enlil. The rest is recorded in the Lugal-E text and the fragmented inscriptions. However, the most striking recount is that of the text "I Sing the Song of the Mother of the Gods", first reported by P. Dhorme in his work "La Souveraine des Dieux (The Queen of the Gods)". It is a poetic text in praise of Ninmah ("the Great Lady") and her role as Mammi

("Mother of the Gods") on both sides of the battle lines. When Ninmah goes to Enlil, Adad; in other words the Hittite God Teshub, is present there, too. While mentioning his ideas, Hathor begs to both of the gods: "Listen to my prayers!" Initially, Adad is very stubborn: "If she wants to bring about a cessation of hostilities, let her arrange and agreement on the basis that the *Enlilites* are about to win."

Marduk (A.Zag), who was at the Great Pyramid at that time, gets overwhelmed by the merciless attacks, and prohibits his followers, who were enduring the besiege at the Giza Complex, from resisting any more: "The enemies made their wives and children fight the war; that God did not raise a hand against Ninurta. The weapons of Kur were covered with soil. A.Zag did not use them." (Figure 4)

Figure 4: The Egyptian Pyramids in the Sumer Tablets

Marduk is resisting in the Great Pyramid

Upon Marduk's approval and Ninmah's efforts the two clans retreat from the Wars of Gods, but the war is not over yet. The Lugal-E text says "Ninurta did not annihilate the scorpion in

Kur." Instead, the enemy gods retreat into the Kur, i.e. the Great Pyramid, by building a shield. Marduk is now all by himself, and safe inside, because due to the safety shield, it is impossible to penetrate from outside to inside. Now Marduk, who is inside the Great Pyramid, and Inanna, Adade and Ninurta, who are outside, are the only ones remaining in the war. It is now high time to make Marduk come out, and hand him over to justice.

Historians working on the fragments of the Mesopotamian tablets scattered in different museums were successful in putting together a text, which was called "Inanna and Ebih" in a book by Samuel Noah Kramer called The Sumerian Mythology. (5) This text tells Inanna's struggle with an evil god hiding inside the Mountain. Not only the inscriptions but another cylinder seal shows that this mountain is the Great Pyramid, and the encounter takes place in Giza. (Figure 5) Hereby, Inanna has her back covered with the support of Enlil (the Bull), and the East clan (the Eagle), and she is fighting a war with a god in an area where the three pyramids are located. The serpents curling on each other, the ankh and the priests in the figure are still symbols of Egypt even in our current times. The serpent is the symbol of the west clan anyway.

Figure 5: The Pyramid War in Sumer

As Inanna keeps challenging Marduk, who is hiding in the Great Pyramid, her fury grows: "My grandfather Enlil has permitted me to go inside The Mountain! Into the heart of The Mountain I shall penetrate... Inside The Mountain, my victory shall establish!" and she continues to attack. But Marduk Works hard; his defense is very strong. A.Zag, that is; "the great serpent" is an epithet given to Marduk by Enlil as far as we can follow from the Ninurta inscriptions. And his hiding place is described as E.Kur, which has walls awesomely reaching the skies; and that is the Great Pyramid. Inanna "ceased not striking the sides of E-Bih and all its corners, even its multitude of raised stones. But inside... the Great Serpent who had gone in, his poison ceased not to spit." Marduk has been described as the scorpion when he starts counter-attacking to defend himself.

Eventually, by the help of Thoth, they break down the defense of Marduk, and enter inside the Great Pyramid. Marduk leaves where he was, which is nowadays called as the queen's chamber, and escapes into the king's chamber. Once inside, he closes the entrance by lowering down the granites in the front room. Ninurta and Inanna, who are now inside, cannot overcome this barrier no matter how much they try. Instead, by placing the granite blocks in the ascending passageway, they enclose Marduk inside. Thus, passage to the upper chambers of the pyramid are closed irrevocably. Marduk is now all alone in the king's chamber. He has no water nor food, but he can breathe through the air channels reaching the north and south faces of the pyramid.

Inanna has given the verdict. Marduk is buried alive in the Great Pyramid, and he is doomed to die there in agony. "The Judging Seven" accept what Inanna does: "The mistress art thou... The fate thou decreest: let it be so!" Assuming that Anu, the King of Nibiru, would go along with the verdict, the judging seven gods then placed the command to Heaven and Earth.

The records on Marduk being buried alive inside the Great Pyramid are also kept on the clay tablets recovered in the ruins of the ancient Assyrian capital cities of Ashur and Nineveh.

The Ashur inscription suggests that, upon the suspension of the judgment, this incident becomes the subject of a ceremonial New Year's play in Babylon in which the suffering and reprieve of the god is reenacted. (6) According to this text, Marduk does not die. This ancient script develops like this: Bel-Marduk is confirmed in The Mountain. The city falls into tumult. In the meantime, Marduk's wife Sarpanit comes into scene. A messenger tells her that Marduk is imprisoned in The Mountain while weeping in front of her. Sarpanit recites an appeal to the two gods who can convince Inanna; Inanna's brother Shamash and her father Sin. She begs to Sin and Shamash. In the meantime, priests come on to the stage appearing in a procession. Suddenly an actor representing Marduk, clothed with shrouds which are dyed with blood, appears on stage, and starts talking: "I am not a sinner! I should not be punished!" He announces that the Judging Seven had reviewed his case and found him not guilty. Another god is found guilty, and he gets slayed. However, Marduk should get a punishment, too, because of his indirect guilt. And this is exile. Sarpanit and everyone else are dumbfounded. Nobody can understand how Marduk can be freed again, if he was imprisoned in a mountain (a tomb) that cannot be unsealed. Inanna and Ninurta had blocked the ascending shaft with three granite Stones. This constitutes a great problem. (Figure 6G) The leaders seek a way to reach the king's chamber in the shortest time possible in order to take out Marduk whose life was in danger.

Seeking out the way to reach Marduk in the fastest possible way Finally, they decide on the best shortcut, and the divine messenger Nusku explains the solution like this:

> A doorway-shaft which the gods will bore;
> Its vortex they will lift off,
> Its abode they shall reenter.
> The door which was barred before him
> At the vortex of the hollowing, into the insides.
> A doorway they shall twistingly bore;
> Getting near, into its midst they will break through.

These lines describing how Marduk would be freed still do not have much meaning for the historians, but for us, they have utmost importance. Thoth, who had designed the pyramids, had created another entrance hidden from everyone else. (Figures 6A to 6F) And he left one point of the tunnel closed on purpose so that it could not be discovered by accident by unwanted people. (Figures 6B to 6D) Thoth, who knows the plan of the pyramid by heart, sees that the shortest and fastest way to save Marduk is to open a linking tunnel of only ten meters through the soft limestone blocks between the B and D cavities. This task would take not days but hours.

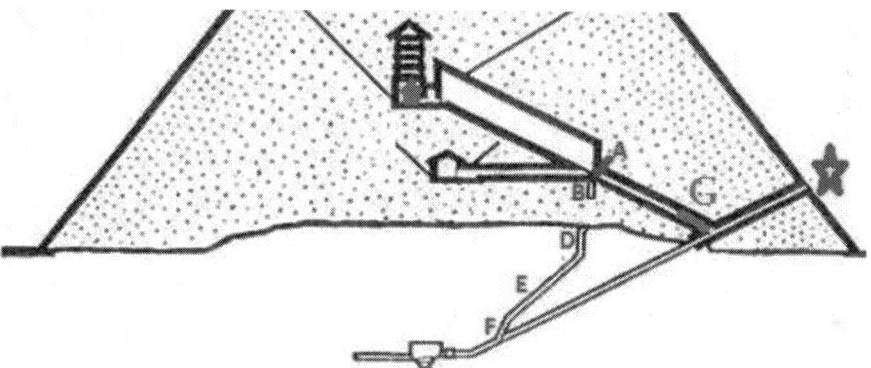

Figure 6: The alternative tunnel in the Great Pyramid

Now let us read into what Nusku told within the light of this information:

A doorway-shaft which the gods will bore;
The rescuers came downwards from the descending tunnel to create an entrance cavity;
Its vortex they will lift off,
They lifted the stone in there, opening up to the F and E cavities.
Its abode they shall reenter.
Climbing quickly up the inclined F and E pieces, they entered into the place where E is connected to the vertical D piece.
The door which was barred before him

The entrance to the cave was sealed with a granite stone. They moved this stone aside.
At the vortex of the hollowing, into the insides.
The rescuers were now at point D, and 10 meters above, but to the side, was the base of the B piece, and the entryway to the grand gallery.
A doorway they shall twistingly bore;
Right away, they broke down the blocks of limestone between B and D. When they reached the connection point with B, they were at a short, horizontal passage; passage A. At this point, they came across with a stone wall in front of them.
Getting near, into its midst they will break through.
Behind this wall were the chambers above the grand gallery, the queen's chamber and upper chambers of the pyramid. To be able to pass into these chambers and passages, the plug-like ramp stone had to be removed, but it could not be moved at all. Well, this is the part when the stone is exploded from the middle.

The hole in mention is still visible there today. Adam Rutherford, who describes it in his book "Pyramidology", suggests that it was created by an explosion: "The hole seemed to be created by an explosion from a phenomenal power from inside the well."

The pieces of the limestone block, which was exploded with a tremendous force, fell through the ascending passage all the way down to where the granite stoppers were. About nine thousand and six hundred years after this, when the Islamic Caliph Al Mamoon discovered the upper rooms for the first time, he saw a thin layer of white dust, which made climbing up the ascending tunnel harder. This was the result of the explosion that had happened thousands of years ago.

The rescuers, who entered the grand gallery from the big opening created by the explosion, enter the king's chamber by moving the three granite blocks in the fore-chamber. Then, they carry the fainted Marduk, and leave from the way they had come in.

The one who enters the Great Pyramid last is Ninurta

So, what happens to the pyramids later on? The Lugal-E text mentions that once Marduk is carried outside, it is Ninurta who enters inside the last. Ninurta is escorted by the Chief Mineralmaster and close soldiers. What could they have possibly found inside? Let us check it in the Lugal-E text.

Ninurta, who entered inside from the hole Thoth had opened, came to the junction of the Grand Gallery, the ascending tunnel, and the queen's chamber. First he came to the queen's chamber passing through the horizontal passage. The axis of this chamber lay exactly on the center line of the pyramid in east-west direction, and there was a tremendous amount of energy flow inside. There was something emitted from the stone in the niche carved on the eastern wall of the chamber. We think that either radiation or some other thing unknown to us was being emitted from a special equipment here. The text hereby mentions what was in the niche as the "Sham" ("Destiny") Stone. Emitting a red radiance in the dark, this stone was the pulsating heart of the pyramid. We learn about the destiny of this and other stones from the Lugal-E text on tablets 10 to 13. By following this text, and interpreting it accurately, we can learn many of the features of the inner structure of the pyramid, and the purposes and functions intended for the stones.

Whatever it was that was emitted from this stone, it had chased Ninurta during the war, and it had caused great damage to Ninurta's black bird. That is why Ninurta "ordered it 'pulled out... be taken apart... and to obliteration be destroyed.'"

Ninurta returns to the junction point of the passages, and stops and looks around in the grand gallery. This gallery glittered in multicolored lights. Its vault was like a rainbow. The many-hued glows were emitted by twenty-seven pairs of diverse crystal stones that were evenly spaced along the whole length of each side of the Gallery. Each crystal stone emitted a different radiance, giving the place its rainbow effect... As Ninurta's priority was the king's chamber, he just passed by them for the time being.

Climbing up from here, he reached the Antechamber. At this point, "the bolt, the bar and the lock" of the Sumerian poem, hermetically sealed off the Grand Chamber: "to foe it is not opened...but only to the live ones". In this chamber, there was an equipment placed in a carved stone chest placed on the north-south axis of the pyramid. This equipment which was named as "The Guide/Radar" was surveying both the Heaven and the Earth. The heart of this radar unit was the GUG Stone.

These stones emitted resonating beams to guide the astronauts, and radiation to protect the structure. Escorted by the chief mineralmaster, Ninurta inspected the array of Stones and instruments inside. The power of the resonance emitted by the GUG Stone was multiplied by five deep partitions built on top of the chamber, and they were spreading outside through two channels. Ninurta ordered this stone destroyed. The GUG Stone was taken out of the niche that day, and smashed to pieces.

Ninurta decided to have the three-lock mechanism at the door to be removed. He was standing right in front of the "Cu, Gasurra, and Sajkal" Stones. He cut the ropes holding them, and destroyed them. (7) Now it was the turn of the twenty seven pairs of mineral stones and crystals in the gallery. Some of the minerals and crystals belonged to the skies, yet some others were like nothing he had seen before. As he kept walking down, he stopped by each one of them, and he determined its destiny. Lugal-E text mentions the names of twenty-two stones. The remaining five names cannot be read due to the broken parts on the tablets. Ninurta decided that the damaged ones would be smashed up and pulverized, others taken away for display, and the rest would be sent to Shamash to put into use. Finally there was the Apex

Stone of the Pyramid, the U ("High as the Sky") Stone. Getting on his flying bird, Ninurta crashed the Apex Stone of the Pyramid. Ninurta announced that he was doing all this not only for his sake, but for the future generations:

"Let the fear of thee" - the Great Pyramid - "be removed from my descendants; let their peace be ordained. Let the children of the Mother Goddess not see this (The Apex Stone) anymore. Let everyone stay away from it." (8)

Following this, everything was taken from the upper chambers of the Great Pyramid; the hole made by Thoth in 8760 BCE was filled in; and the pyramid was sealed all empty and deserted, and without an apex stone.

In the following years, many people have gone in and out of the pyramids. Although Herodotus and Strabo mention in their works that there was entry into the Great Pyramid, it is only about entering the lower chambers. As the existence of the upper chambers were never known, all the other pyramids ordered to be built by the pharaohs in Egypt were constructed without the upper chambers. Those who want to have more information on the Pyramid Wars can read Zecharia Sitchin's book "The Wars of Gods and Men". Ninurta's victory at the Giza Pyramids has long-lived on a cylinder seal. (Figure 7)

Figure 7: Ninurta's Eagle and the Pyramids

Ninurta's symbol the double-headed eagle gives inspiration even today to the states related to the east clan.

On the other hand, the Great Pyramid, which was sealed by Ninurta, and the destroyed Apex Stone still survives as the symbol of the states of the west clan and organizations such as the Illuminati.

Sources:

1. https://atlantisrisingmagazine.com/article/more-evidence-uncovered-in-howard-vyse-pyramid-fraud/
2. http://www.history.com/news/piltdown-man-hoax-100-years-ago
3. SİTCHİN, Zecharia, 12. Gezegen, Ruh ve Madde Yayınları, İstanbul, 2001,
4. Lugal-E http://etcsl.orinst.ox.ac.uk/section1/tr162.htm
5. Inanna and Ebih http://etcsl.orinst.ox.ac.uk/cgi-bin/etcsl.cgi?text=t.1.3.2&charenc=j#
6. Bu metinler Heinrich Zimmern tarafından Berlin Müzesinde okunmuş ve 1921 Eylül'ünde bilim dünyasına sunulduğunda büyük bir karışıklığa neden olmuştur. Stephen Langdon metne "Bel-Marduk'un ölümü ve yeniden dirilişi" adını vererek yayınlamıştır.
7. Cu taşı: dikey taş gasurra taşı: korkutucu, açılabilir olan sajkal taşı: önde duran sağlam taş
8. SİTCHİN, Zecharia, Tanrıların ve İnsanların Savaşları, Ruh ve Madde Yayınları, İstanbul, 2005, S.215

SECTION 9

AN ALTERNATIVE HISTORY OF SIX THOUSAND YEARS

If you would be a real seeker after truth, it is necessary that at least once in your life you doubt, as far as possible, all things.

René Descartes

From the War at the Great Pyramid to the Illuminati

The upper chambers of the Great Pyramid had been waiting all sealed since the victory of Ninurta until the Islamic Caliph Al Mamoon discovered them in 820 ACE. Based on the accounts of the Arabic historians, the Caliph had formed an army consisting of masons, ironsmiths and engineers in order to enter the pyramid. Al Mamoon's men initially dug out random tunnels in the walls forcing an entryway. Later on, they heated and cooled down the stones over and over again until they cracked. Eventually, they were successful in getting inside the pyramid.

The men clear their way to the original passage by breaking the stones, and then they walk up the passage, and reach at the entryway which did not let them in from the outside. After opening the door to outside from here, they descend through the passage to the well described by the famous geographer Strabon. Finding nothing but this passage and the well, the soldiers were just about to give up and return. Right at that moment, they hear the noise of a falling stone. They look for it, and find it. This stone, which was in a peculiar triangular shape, was sandwiched in, in order to hide a big, rectangular granite slice located in accordance with the angle of the descending passage. Obviously, no one has seen these granite stones before. Thinking that these granite stones were hiding an unknown secret chamber, the men announce the finding to the Caliph right away. Upon the order of the Caliph, they give a big struggle on the granite block. However, they understand that it was only one of a series of big granite blocks with limestone blocks in between them. Then, an alternative solution is thought of: If the granite stones could not be dug out, then, the limestone blocks right next to them would be dug out. In this way, they would go beyond the granite stones. Well, these granite blocks happened to be the *very* blocks Inanna and Ninurta had used while sealing the passage above the ascending passage in 8760 BCE. Consequent to this process, Marduk had been imprisoned alive inside the pyramid, and passage to the upper chambers of the pyramid had been irrevocably closed.

Following many days of excavations, Al Mamoon's men enter the ascending passage, and begin climbing. At the end of this ascent, they come across with another passage which lines up horizontally. When they enter and follow this horizontal passage, they reach into a square-shaped chamber with a pointy-ridged roof. And this is what is known as the Queen's Chamber nowadays. However, this chamber was barren and empty. Al Mamoon, who did not find what he expected in the Queen's Chamber, orders his men to climb up even higher in the passage. The men come across with a big gallery opening upwards from the ascending passage. The gallery is 45,7 meters long, and it ascends by a 26 degrees angle. The ground is caved-in, and there are two ramps extending all the length of the gallery. On each ramp there are rectangular holes located at steady distances, and facing each other. The niches, which were cut out properly, and which are empty now, help this climb. These are the niches where the twenty-seven crystal stones mentioned in the Lugal-E Text stood. This task got more difficult due to the white layer of dust sprinkled on the floor, and the inclined surfaces of the gallery. Well, this white layer of dust is the dust coming from the exploded stone during the efforts to reach Marduk. They reach the big step, which was elevated from the upper edge of the gallery, so that it would be at the same level with the floor of the antechamber, and they enter the antechamber. When they enter the antechamber, they see that the portcullis block was not in its place. The antechamber forms a big piece of stone, a straight platform at the top edge of the gallery. A short, about one meter wide, and low corridor at the same level with the platform is connected to the antechamber, which is built in an extremely complex manner. This part is equipped with a mechanism, which provides the three strong granite walls that block the gallery vertically and hinder going any further, to move down with a simple movement. From here, they advance —once again by crawling on the ground— into the vaulted chamber above, which would be later called the King's Chamber. Unluckily, this chamber was empty like before except for the chest with the carvings, which was also empty. The chest, which has a meticulous craftsmanship, has a lid, and pits for the upper part. The dimensions reflect an advanced mathematical knowledge.

The men of Al Mamoon do not find anything inside the Great Pyramid. Later, they find the well Thoth had sunk in a hurry to save Marduk, and they break the stones at its bottom reaching at the initial passage. As witnessed by the Arab historians, Al Mamoon found nothing but void inside.

Since that day many people have entered the Great Pyramid, and have seen the upper chambers. The Pyramids and the Sphinx which are regarded as a tourism center, but they have much deeper meanings than known. The apex stone, which was destroyed in the War of Pyramids that was won by the east clan, has been the most important symbol for the West clan, and it has been placed on top of the Illuminati. It points to the power and leadership of the Egyptian symbols Enki/Ptah and Marduk/Ra covered in a sauce of Illuminati, which we come across everywhere nowadays, in the Age of Aquarius that will last for 2160 years.

Furthermore, we believe that the governing of all the secret western organizations that rule the world in the new age belongs to Marduk/Ra. We suggest that the eye, which sees everything, placed on the apex stone depicted separate from the pyramid —one of the symbols of USA today— is Marduk's eye. (Figure 1)

The Latin term "Annuit Coeptis" in this symbol is translated as "God approves the beginnings", but it is quite questionable. It is also used in reference to "He favors endeavors or to smile upon, to favor" (1) The other motto as part of the symbol is "Novus Ordo Seclorum" which has been translated as "the new order of the ages/history". These two mottos combined indicates to the meaning as "God approves the new order of history." However, this meaning seems more plausible: "Anu approves the new age (The Age of Aquarius)." Keeping in mind that Marduk/Ra is the leader of the Age of the Ram, the Ionic style corrugated ends of the ribbon on the left and right sides of the motto, which look like ram's horns, take a different meaning. Furthermore, the number right below; i.e. MDCCLXXVI stands for 1776, which is both the date of the foundation of USA, and that of the Illuminati Organization.

Figure 1: Symbol of USA

When we analyze the last names of the two families, which allegedly rule the world, we see an interesting Picture. Rockefeller means RA-KA Pharaohs, and Rothschild stands for (RA-KA-M = RA-KAM <KAM = shield> = RA-Shield) the Shield of RA.

These names explicitly display the connection of the current subliminal war with Ra. (2) It is a prevailing belief that the Rockefellers and the Rothschild are designing the world today. What is unknown is the Anunnaki dimension of the phenomena. Throughout history a phenomenon like two individuals ruling the entire world has not been experienced. Consequently, it is not logical for people who live for hundred years to establish organizations that last for thousands of years. It is more plausible that the Anunnaki established these organizations, and they rule the world through the people/families they choose. Marduk, who is the leader of the Age of the Ram for the Anunnaki, is one of the three Anunnaki; Isis, El and Marduk, who had a duty in the foundation of Is-Ra-El. Within the light of all this information, now we can begin our journey of six thousand years of alternative history.

The Sumer Civilization

Since the times Enki divided the Great Precession Period of 25920 years into twelve horoscopes of exactly 2160 years each, the expectation for each new age to be the "Golden Age" has been around. The term "Golden Age" has been the popular talk of town at every turn of a new age in the past just as it is today. Humanity has known its place and role since the times when priesthood and kingdom have been granted to them. Based on this, for humans the deities (the Anunnaki) are masters to be worshipped and respected. There is a definite hierarchy, predetermined ceremonies, and holy days. For thousands of years, the deities have been attending the well-being and faith of humanity, but they have always stayed at a distance. Only the high priests were able to get closer to them on certain days, and the kings have communicated with them through visions, dreams, and prophecies. Every nation has thought that the new age will be bringing them the "Golden Age" as long as they adhered and applied the rules of its own god strictly. And this has caused the religion wars; thousands of people have died and killed for their deities and rituals.

So, these changes of ages were made according to what? Who was managing the transformation of ages? Why were the transitions always happening through wars? In order to be able to give the answers to these questions, we need to go on a journey from the past to today.

In the period between the Great Flood at around 11 thousand BCE to 4 thousand BCE, the earthlings reproduced and multiplied in numbers, going from mountainous regions to dried-out plains using the information given by the Anunnaki, and by means of their domestications in agriculture and livestock breeding.

By 4 thousands BCE, the civilized people descending from the race of Ham, Shem and Japheth —the sons of Ziusudra (Noah)—, had already mixed with the seed of the Anunnaki; the Igigi (the Watchers-fallen angels), who had married with the earthlings, had produced grandchildren spreading all around; and on the American continent, the relatives of Ka-in had achieved completing great works.

During this period, as the population of humanity increased rapidly, and the Anunnaki, who had come from Nibiru, decreased in numbers, the problem of how they would stay holy in the eyes of humanity arose. They must have contemplated on how they could secure submission and service from the humans, who outnumbered them. Right at this point, the great deity Anu decides to come to Earth one more time from his adobe in heavens; from Nibiru. When Anu's space ship arrives on Earth, the Anunnaki ascend and glide in their vehicles, and guide him so that he can land safely at the spaceport in the Sinai Peninsula. After this visit, during which all the Anunnaki were present, Anu and Antu sit at their throne at the palace built in Uruk. Following lengthy conversations and discussions, Anu changes the direction of history by taking a very crucial decision: "We are only messengers for the earthlings on Earth. The Earth belongs to the earthlings; we are here to protect and advance them!"

Upon this decision, the period for the foundation of human civilizations begins. Buildings similar to those in the Kingdom of Nibiru are designed so that information could flow systematically to humanity. From that time on, the crown and the scepter would be given to a selected human being, and the Anunnaki would tell their words to humanity through that person. In this way, people would be trained in their working and manual skills. Besides, clergy was formed in holy regions so that humans could serve and worship the Anunnaki as their "holy masters". From that time on, the secret knowledge would be taught, and civilization would be transferred to humanity.

The Anunnaki decide to establish four regions, one of which would be prohibited for humans, and the other three handed over to them. The first region, in which Enlil and his sons reigned over, and which we know as Sumer, was established in Aden in 3800s BCE. Then, the second region, which would become the Egyptian Civilization, is settled on the two poverty lands through the efforts of Enki and his sons. The third region, not to be mixed with the other two, was established in a far land, and it was given under the administration of Inanna; this is the civilization we know as Indus. And the fourth region, which was appointed holy only for the Anunnaki, is the Sinai Peninsula where the spaceport in the leadership of Ninmah is located.

A new time calculation method is invented in order to immortalize this visit of Anu. From that time on, whatever happens on Earth would be calculated in terms of Earth years, not in Nibirun shars. The counting of the years is initiated at the Age of the Bull dedicated to Enlil (3760 BCE), and this first calendar has been named as the Nippur Calendar.

Also, during this period, the Anunnaki have been honored with rank numbers by humans in order to acknowledge their heritage rights. Heavenly Anu gets number sixty. Enlil is given rank number fifty. Enki gets forty; Sin thirty; and his son and consecutive Utu/Shamash receives twenty as his rank number. Rank number ten has been allocated to the sons of the other

Anunnaki leaders. Rank numbers ending with five have been distributed among the Anunnaki women and wives. We think that these ranks have to do with jurisdiction and frequency.

Sumer Civilization being established by Enlil

Humans begin founding their cities by the help of the Anunnaki once they learn how to make bricks from mud. At locations where once only the Anunnaki cities existed, now both theirs and the earthlings' cities rose. In the new cities ziggurats were built dedicated to the prominent Anunnaki. These ziggurats were attributed holiness, and consequently they were turned into temples. The Anunnaki were served in these temples as the "Holy Masters". In the later years, they become places of worship for the polytheistic religions.

Once the temple-adobe of Enki and Enlil were completed in Eridu and Nippur, the district of Girsu was constructed for Ninurta in Lagash. Ninurta's and his wife Bau's temple-adobe was called Eninnu. At the location where the Sippar stood before the Great Flood, to the peak of the muddy soil, a new Sippar was erected, and for Utu (Shamash) and his wife Aya, the Ebabbar Temple was built as an adobe. The laws of justice humans would be bound by were proclaimed from here. New areas were chosen for each area where the old plans could not be implemented due to sand, eluvium and mud. In a place not so far from Shurubak; in Adab, a new center was established for Ninmah. A city called Ur, which had perpendicular streets, channels and docks, was established for Sin (Nannar). Ishkur (Adad) had returned to the mountainous lands in the North. His adobe was called the House of Seven Storms. Adad would be renamed as Teshup in the future, and his adobe would be named Hittite. Inanna remained in Uruk where she lived in the house Anu gave as a present. (Figure 2)

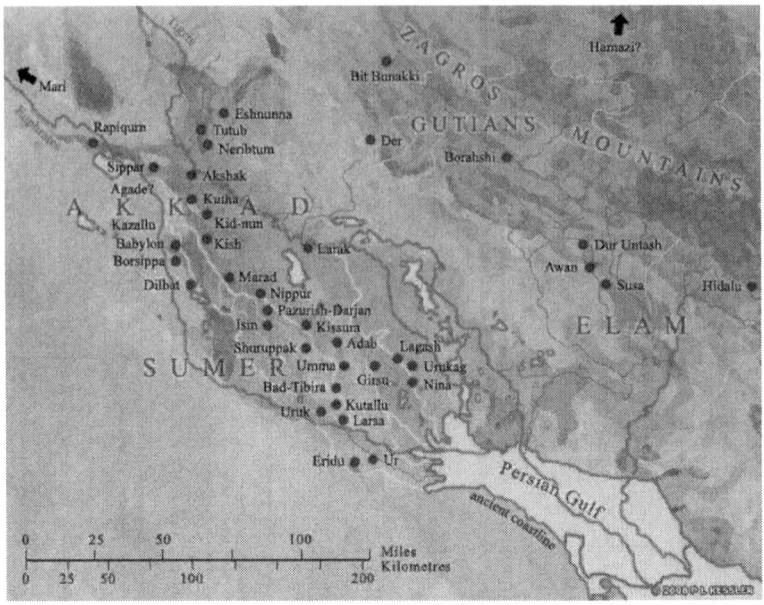

(Figure 2): Sumer Cities

In the first region called Sumer, the Anunnaki teach the handicrafts and manual skills. Not before long, the fields were irrigated; boats were sailing in the channels and the river; pens and granaries were filled and brimming over; and abundance and fertility was everywhere. The first region is named as Ki-Engi, which means "The Land of the Holy Guards". Then, they decide to let the black-headed nation have their own city, and in the city of Kish, the first human kingdom is initiated. According to the Sumerian Kings List, the first postdiluvian city to have kings is Kish.

Egyptian Civilization being established by Enki (Ptah)

The establishment of the Egyptian Civilization was deferred a little. Egypt's systematic rise starts in 3150 BCE. The reason for this is Ningishzida being in charge of the governing when Marduk, the son of Enki, returns to Egypt. The country was in chaos because of these two brothers who had been struggling with each other for three hundred and fifty years. Eventually,

Ningishzida abandons Egypt following Enki's intermediary efforts, and he goes to the continent of America together with some of his followers. This migration of Ningishzida in 3113 BCE has been considered the beginning of the Mayan Calendar. Ningishzida was called as "Quetzalcoatl" meaning "Feathered Serpent" in this half of the world. The civilization he established would later be called "Olmec", and the advanced cities founded by the Olmec would be inherited by the regions' indigenous societies; the Incas, Mayans, and the Aztecs. On the other side, in Egypt, under Marduk's governance, the second region gets established. The second region gets mentioned in the history records as the Country of Magan. In the new language of Egypt, the name of the country becomes Hem-Ta, and the Anunnaki become the "Neteru"; the Guards.

Marduk was worshipped as "Ra – The Bright One"; Enki was respected as "Ptah – The Developer"; and Ningishzida was called as "Thoth – The Divine Measurer". Thoth's face has been scraped off of the Sphinx; and instead, Osiris' face has been placed. Ra forces the public to count sixty by sixty instead of ten by ten; divided the year in ten, and chooses the Sun Calendar instead of the Moon Calendar.

While under Tehuti's mastership, the old North City and South City were rebuilt, Marduk/Ra combined these two, and made them into one country. He appointed a half-god king; Mena, who was born from the marriage of an Anunnaki and an earthling, and then, he named the city founded in 3050 BCE as Mena-Nefer. Ptah gives Ra all types of ME (codes loaded with knowledge) so that the second region can benefit. Because Ptah regulated the water flow of the Nile, bounty arrived at the fertile lands; and the people and cattle rapidly multiplied in numbers.

The Council of Twelve makes an Agreement based on the Constellation

The constellation is actually symbolically shared between the twelve Anunnaki leaders. However, after the calendar calculation, this sharing was directly related to leadership of the world. Marduk claims right on the world leadership pointing out to the approaching Ram Precession. By the help of his son Nabu, he sends messengers to the lands which do not belong to him, and announces to the whole world that his age has arrived. He gets the nations accept his leadership by telling them that the Ram Constellation Precession has begun. Initially, Enlil and his clan do not give much regard to this, but, later on, they come to understand the seriousness of the matter, and take certain precautions. One of the precautions taken is Thot teaching sky observation to the public. Thoth designed stone buildings using his wisdom, and Ninurta and Ishkur helped him. This trio taught human beings how they would observe the near and far skies, and showed them that the sun was still rising in the Bull Constellation. Stone calendars such as Stonehenge and Golan Heights display the precession of ages on March 21 even today. That is why, at many of the stone circles, rituals for spring equinox are held.

No one knows today whether we are still in Pisces or passed into the Age of Aquarius because we do not exactly know the limits appointed to this twelve constellations by Enki, and we can only make guesses on this point. We are guessing that the Age of Ram Precession had started right on the date Marduk says, that is; on 2260 BCE. The Age of Pisces, which started on 100 BCE will come to an end on 2060 ACE, and on this date the Age of Aquarius will begin. The seventy-two year period corresponding to the last one degree of the Age of Pisces began in 1989, and we have been lucky enough to be able to watch the leap of humanity, and its reach at information in the period after this date, when big changes started occurring. Let us remind right away: The International Astronomical Union updated the constellations as eighty-eight in 1922 by the contributions of Henry Norris Russell, and those entering the Zodiac were increased to thirteen. And in 1930, the Belgian astronomer Eugene Delporte prepared a map determining the definite limits of these constellations with each other. (3) What we are referring to is not this man-made collocation, but the limits we inherit from the ancient times.

Let us move on with our subject. As result of the disagreements and wars among the Anunnaki, there were nuclear attacks. The spaceport at the Sinai Peninsula, and five Dead Sea cities belonging to the followers of Marduk were affected by the nuclear attacks, and the Sumer cities were extinguished by 2023 BCE due to the nuclear cloud. Consequently, Sumerians become the first migrant nation in history, and take civilization to wherever they go. Enlil's transfer of leadership to Marduk happens right after this bitter incident, but it is not accepted right away. Following this announcement, a chaotic atmosphere presided over the region for exactly a period of one hundred and ten years. Later on, all the Anunnaki unite through an agreement.

Marduk's leadership is accepted in the Age of Ram

In 1890 BCE, a ceremony was held in Babylon in which all the Anunnaki participated. The ceremony for Marduk's ascension to the throne as the "King of Gods" takes place in the presence of fifty great deities, seven deities of destiny, and hundreds of Anunnaki from all ranks. The great Anunnaki prince; Enlil, who had been the leader of the world for 400 thousand years, passes the leadership on to Marduk/ Amon Ra, thinking that this is what faith desires, following the disaster in 2023 BCE. Enlil leaves the Arch, which is his own weapon, in front of Marduk as a symbolic gesture. Consecutively, as the next phase of the transfer of Enlil's powers to Marduk, Marduk was given fifty ranks. These fifty names were read out loud one after another. These started with the birth name of Marduk, followed by the rest of the titles, and ended with him turning into the most superior planetary god.

In the Enuma Elish Text, Enki announces at the conclusion of the ceremony: "All of my instructions are embodied in the fifty names. In those, all the rites have been combined." Enki writes these down using his very own hands, and adds that he will be preserving them for the generations to come. Then, he orders this piece of writing to be stored at the Esagil Temple, which would be built in Babylon for Marduk by the gods. There, the secret knowledge would be safeguarded by a line of priestly initiates, passed from father to son: "Let them be kept there; let the elder explain them; let the wise and knowing father impart to the son."

As expressed by Enki, the fifty names "which combine all that should be known" are the authorities of Marduk, who is the leader of the Council of Twelve for the period. Besides, these titles are considered to be divine secrets, and they are only explained to the priests and initiates "in Sumer: la mudu".

During this ceremony, in which Marduk's leadership is announced, the Council of Twelve is updated, and decision is taken that leadership would be undertaken consecutively by the Enki and Enlil clans. The leadership for the Age of Ram will be in the hands of Marduk. As the Age of Pisces begins, the Council of Twelve will be updated, and the leadership will be handed over to Sin (Nannar) from the Enlil Clan. In the same way, the Age of Aquarius will pass by with Enki, and the Council of Twelve will be determined all over again. As long as the flow of gold to Nibiru continues without interruption, and the human beings do not revolt against the Anunnaki, these leaders will be the sole authority on Earth for the order on Earth, and there will be no interference with their decisions. In the distribution of authority and frequency numbers, the decisions which can be taken by the leader without consenting to the council are determined based on the fifty names given in Babylon. In other words, based on the authority granted to him, the Anunnaki leader will keep the population of humans on Earth in balance; maintain the order of the world, and organize the transportation of gold to Nibiru. The leadership will change hands from the Enlil clan to the Enki clan consecutively, and vice versa.

The most important event at the beginning of the Age of Ram is, without a doubt, the collapse of the Sumer Civilization. Majority of the Sumer nation was extinguished during this transition, and the remaining ones were scattered at various places as refugees of the first great exile of history. In the Mesopotamian texts the word Munnabtutu, used in the sense of refugee, stands for "those who escape demolition". Based on our own experience, it is also possible to translate this as "the routed persons". The Sumerians carry civilizations with them to the places they go. The Sumerian physicians, astronomers, architects, sculptors, seal cutters, and inscribers become the teachers of the nations in the other countries.

The Sumer refugees were taken under the auspices of all the countries surrounding Mesopotamia, and they became the catalyzers of the hosting countries in turning into modern and developing states. Some had traveled even further. According to some scientists, in 2000 BCE; and according to others, in 1800 BCE, "a sudden mysterious change" had occurred in China. Originally, this country was consisting of primitive villages. Then, all of a sudden, it turned into a country with cities circled in walls, in which the emperors had bronze weapons, horse-drawn chariots, and the knowledge of writing. In a short while following the collapse of Sumer, sudden new civilizations had popped up in China. Although it is not very clear, it is thought that writing had started simultaneously with the kingdom of Shang Dynasty. This script is mono-syllable, and it uses figure-like characters. As is known, these are the defining characteristics of the Sumerian script. Based on this, the familiar Chinese characters have been converted to somewhat cuneiform. The similarities between Chinese and Sumer inscriptions, and the relation between these two writing systems are the subjects of many scientific thesis. Especially the astonishing resemblance of the Sumerian hieroglyphs to the old forms of Chinese script has been proved in a definite way.

Shintoism religion of the Japanese is based on the belief that their emperor is the son of the Sun god and goddess. The

Japanese keep the Sun Goddess Amaterasu alive with the red circle on their flag. We consider Amaterasu to be identical with the Sumerian Sun God. Therefore, Shintoism might have been imported from Sumer to Japan.

If we follow the traces in linguistics, we can say that some of the Sumer refugees passed into the West, to Europe via the Caucasians and Anatolia. This hypothesis of ours may explain the language of the Georgian nation being akin to Sumerian. Furthermore, because the Finnish language has no resemblance to anything else but to Sumerian, we can concur that they went through the same route all the way to Finland. We can also infer that another branch went to Hungary from Anatolia by looking at the similarities between the Hungarian and Sumerian languages. The permanent and deep-rooted belief of the Hungarians that their language originates in nothing but Sumerian is perhaps further than a belief.

One of the most obvious evidences supporting that the Sumer refugees followed this route is a historic artifact from the town of Dacia in Romania. At the junction where River Danube meets with the Black Sea, which in the Sarmizegetusa region of Dacia, there is a structure called as the "Calendar Temples" by the researchers. This ancient monument, also referred to as the "Stonehenge by the Black Sea coast" reflects several characteristics of Sumer first of which is the Sumer sexagenarian mathematical system. At a time when the Sumer Civilization was just getting recognized in the world, during his inauguration speech at the Assembly, Ataturk said that we are the grandchildren of Japheth, and right after this he founded Sumerbank. This could be a little more than just a simple coincidence.

There is no room for discussion on the fact that the Sumer Civilization went through a great collapse in 2023 BCE. Those surviving this collapse fell into the status of refugees, and took civilization to wherever they went. Although four thousand years has passed since then, it is still possible to see all the traces everywhere. And what is bitter is that these lands, in which the Sumerians were refugees back then, have never given peace to any nation, and have always forced the people, who have settled on them, to be constant refugees. This destiny has not changed in the current times, and perhaps it will not change in the future, either.

Towards the Age of Pisces

The Age of Ram is the age of polytheistic religions. As the Age of Ram reaches to the end, and the beginning of the Age of Pisces approaches, Sin's efforts in transiting to monotheistic order made themselves felt even five hundred years beforehand. Contrary to the Age of Ram, during the Age of Pisces, the goal was to transit from polytheistic religions to monotheistic ones. This time, the destruction of the Mesopotamian cities alone would not suffice, because not only the population of humans had increased but they had also spread into several regions. Humanity has been tested through bloody wars especially within the last five hundred years. The children of Israel singled down the "Elohim"—plural deities— mentioned in the early books of the Bible, and adopted Yahweh as the one and only god towards the final books. This might have caused many confusions in the expressions, but in fact they point to very important hints. According to us, Sin, who is the second son of Enlil, is no other than Yahweh.

The Persians were very kind to the Jews when they demolished Babylon; they let them found the State of Israel; and they supported the building of the Temple. In this way, the Temple of Solomon was built for a second time in the leadership of Ezra and Nehemia. While the Persian Armies' efforts in trying to make everyone in the region accept Zoroastrianism continues on one side, Shamash and Inanna —the children of Sin— try to form their own civilization through bloodline in Central Asia. It is said that a few genetic tricks are applied on the children born from the marriages of these two siblings with the humans, and in this way, special races are derived carrying the blood of Anunnaki, which is deemed noble by themselves.

In this respect, the lines of the Epic of Descent (ancestry, lineage) is very meaningful for us.

"We do not have the original and full texts of the Turkish epics from the pre-Islamic period. The scattered information in the Chinese, Arabic, Iranian, Turkish, etc... sources indicate that the tradition of epics in the history of Turks go way back. (1) The Epic of Descent derived from the Chinese sources, which is accepted as one of these epics..."

> *Tola and Selenga come together at a junction,*
> *And join into one river,*
> *At the junction, an island is seen.*
> *In the middle of the island, a hill reaches the skies.*
> *At the top of the hill, a beech tree has turned green.*
> *Days pass by, times go by, a light appears,*
> *When the light descends from the skies,*
> *The beech tree fills up with heavenly light.*
> *Whenever the sun sets down,*
> *And the light descends from the skies,*
> *Sounds emerge from the beech tree,*
> *And everyone listens to music.*
> *Hearing this, all the Uyghur get puzzled,*
> *Those who see this lose their minds.*
> *For ten months and ten days the beech tree stays surrounded in light.*
> *One day, right at dawn, the beech tree suddenly splits open.*
> *Five beautiful children emerge,*
> *From the middle of the beech tree,*
> *They dazzle the eyes,*
> *People stare behind them.*
> *Days pass by, times go by, all grow huge,*
> *The young one; Bogu Khan, becomes the Khan of the Uyghur.*

Let us not continue without mentioning these parts in which Herodotus talks about the agricultural technology thrown at the Scythians from the Anunnaki's spacecrafts:

"In the reign of these there came down from heaven certain things wrought of gold, a plough, a yoke, a battle-axe, and a cup, and fell in the Scythian land: and first the eldest saw and came near them, desiring to take them, but the gold blazed with fire when he approached it: then when he had gone away from it, the second approached, and again it did the same thing. These then the gold repelled by blazing with fire; but when the third and youngest came up to it, the flame was quenched, and he carried them to his own house. The elder brothers then, acknowledging the significance of this thing), delivered the whole of the kingly power to the youngest." (4)

In the meantime, the Greek philosophers begin seeking the mono god in the nature. These special people, who belonged to polytheistic religions, have in fact noticed the existence of a great power, which had created the universe, and thus, they begin questioning. The Anunnaki begin transferring their "monotheistic" belief to the humans with small variations. We can even say that the difference between the Anunnaki and the universal creator has been revealed directly by Sin (Nannar) himself through intuition he implements on special people he chooses.

The Age of Pisces begins

The time has arrived to put the humans into monotheistic belief. The Roman Empire has been the greatest service for achieving this. Due to this forced process, humanity has endured great massacres. We learn from the period's sources that an expectation for the "Golden Age" was articulated in abundance in this period. Unluckily, Rome has brought the "Golden Age" with bloodshed. While Mesopotamia has been turned into a lake of blood over and over again, Europe and Asia gets their share, too. Keeping aside a great portion of Asia in exemption, the establishment of monotheistic religions have been successful.

The most definitive events of the Age of Pisces are the Roman State's evolvement into an empire and the birth of Jesus. Because this Golden Age was the age of a monotheistic religion, there was no need for the information given prior to this. The libraries burnt down during this period has been the best supporting evidence for this.

The polytheistic nations and religions have initiated the new age by resetting time upon the introduction of the new Gregorian calendar.

The balance of the human population has been provided through earthquakes, wars, and especially epidemics. Besides, the most prominent feature of this period is the religion wars. For the last two thousand years, the eastern empires have been leading. It can be said that this age, on which the Turks have a stamp, has been the "Age of the Turks".

Towards the Age of Aquarius

In 1453, when Mehmet the Conqueror conquered Istanbul, the western world had hit the bottom, because the best of technology, the best scientists, the best statesmen, and the best army was those of the Turks'. A little research on Mehmet the Conqueror shows him to be the most intellectual leader of his period. He was speaking Arabic, Persian, Latin, Italian, Greek, and Serbian at the age of nineteen; he was interested in philosophy; he was reading the Greek calligraphy, and gathering all the philosophers around him. He had a grip on mathematics at the level of trigonometry, and he loved reading the star catalogues prepared by Astronomer Ali Kuscu in 1438, and the theories of mathematics. (5)

After this date, which I consider as the bottom point for the Europeans, power starts shifting suddenly to the West. Historians have made different explanations on this matter, and have evaluated it from different angles. They have questioned the power changing hands within the frame of causal links or chain of events caused by phenomena. However, we find these explanations insufficient. After all, it is an interesting change; while the East, where science was functional for the last two thousand years, collapsed, the West advanced in sciences all of a sudden. The Western world, which had not shown a sign of advancement in the name of civilization for the last two thousand years, suddenly appears on the stage of history, just like the Sumerians.

We think that this date is the beginning date of the Anunnaki hand-over process. And the preparations for the Age of Aquarius have begun at these times when we are approaching the end of the Age of Pisces. The developments in the last five hundred years are like the messengers of the new age awaiting for us. Before we explain why we think in this way, let us determine one last thing about the Conquest, and then, look into the developments after this one by one until current times:

At the time when Mehmet the Conqueror conquered Istanbul, what the Europeans knew about history was only constituted of what Herodotus, and other Greek historians had written, and the books of the Torah. What they knew about astronomy comprised of nothing but the existence of the Sun, the Earth, the Moon, and four other planets, and that all these celestial objects turned around the Earth. Courts of Inquisition were burning alive those, who dealt with energy stuff, as witches; and bringing those, who thought scientifically, to trial. The conditions of arts and aesthetics were pitiful.

Well, it is in such an environment that as if someone pushes the button, and the direction of the Western world, which has no preparations whatsoever, changes suddenly.

What has changed in the last five hundred years?

We consider Christopher Colombus' reaching at the riches of the continent of America as a milestone, because carrying the gold of the temples of Maya, Aztec, and Inca by ships made the Europeans rich through a shortcut. As a consequence of the Renaissance and the Reformation, which broke out simultaneously with the overflow of gold, people's perception of religion and arts started changing. We do not know how both Explorer Christopher Columbus and Martin Luther, the Father of the Reformation, reached at the knowledge that would form the foundation of the change. However, once we look at the developments following this date, we can make an inference that they were led by the Anunnaki. Taking into account especially Admiral Byrd's words to Enki as "Dark ages came, but they ended five hundred years ago."; if the 15th century is the end of the dark ages for the Western World, perhaps Enki had a finger in it. We are suggesting that there was an Anunnaki interference in the majority of the early discoveries playing a role in the changes based on knowledge in the West.

Astronomical and Scientific Developments

Right after the geographical discoveries, and the movements of Renaissance and Reformation, the firsts in astronomy are experienced. The "Copernicus Heliocentrism" published in 1543 shortly before Copernicus' death is said to be the starting point of the modern astronomical and scientific developments, and it is deemed as a milestone in the history of science. (6)

Galileo is given such nicknames as "father of observational astronomy", "the father of modern physics" and "the father of science", because he proves the phases of Venus by observing through the telescope; discovers the greatest four satellites of Jupiter; and observes and analyzes the Sun spots. (7)

Kepler is recognized as part of the reforms of the century with Kepler's Law of Planetary Motion, which he discovered in conjunction with his works "Astronomia Nova", "Harmonices Mundi" and "Epitome of Copernican Astronomy". Besides, these works support Isaac Newton's Law of Universal Gravity. Kepler is the first person to think that there is a system of laws determining the movements of the planets around the Sun. (8) Isaac Newton, who was born in 1643, brings a break through into the science of physics, and he is deemed as one of the most influential persons in history by the scientists. Newton not only thought of the law of gravity, but he happens to be the first person to bring an acceptable explanation for the movements of the planets. Furthermore, and more importantly, he has invented the reflecting telescope. (9)

After about one hundred and fifty years, in 1781, Sir William Herschel discovers the planet of Uranus by the use of telescopes. During this period, people were awed at the idea that there were planets beyond Saturn in the Solar System. In humanity's space adventure, after the invention of telescope, the second big step was the discovery of Uranus. (10)

The science field of astronomy gains a new dimension with the discovery of Neptune on September 23, 1846, because Neptune is the first and only planet discovered through mathematical predictions before implementing experimental observations. In 1840, through the use of classic mechanics, Urban Le Verrier analyzes the discrepancies with Uranus' orbit, and he predicts the position of Neptune, which was an undiscovered planet at the time. In 1894, Percival Lowell, an affluent Bostonian founds the Lowell Observatory. Thus, observing Neptune becomes possible, and Lowell for the first time thinks Neptune to be "Planet X". However, upon Neptune becoming definite, he suspects the existence of a new X planet. (11) This suspicion of Lowell causes the Harvard College Observatory to announce on March 13, 1930 that Pluton has been identified. (12)

In 1978, James W. Christie, Robert S. Harrington and Thomas C. Van Flandern from the US Naval Observatory in Washington think that there is a Planet X beyond Pluton, due to the discrepancies on the orbits of Uranus, Neptune and Pluton, as well as on the estimated route of the Halley Comet. Thus they begin their research, which soon turns into the searches of the US and USSR Space Agencies. As a result, in 1983, the Infrared Astronomical Satellite; IRAS is launched into space. IRAS, which was expected to survey all the skies, and research into the near surrounding of the Solar System using infrared technology, observes two hundred and fifty thousand celestial objects before it runs out of liquid helium, which cools down IRAS. The information on IRAS recording a mysterious object resembling a comet circulates in the press at that time, and gets discussed a lot, but the official authorities deny it. According to us, this recorded mystery object is Planet X; in other words, Sumer's Nibiru.

The truth about Nibiru has been kept in secrecy since that day. The announcement made in 2016; "This time, we are sure about Planet X." by Mike Brown and Konstantin Baytgin, two NASA scientists, aims for preparing us, (13) Because very soon, this invasionist planet approaching will be determined by the telescopes of all nations. Just like in the early years of Neptune's discovery, it is claimed that visual contact has not been established yet, but, on the other hand, the intensive effects it has on our Earth are being observed. Besides the increasing meteorite traffic in the last decade, intriguing hurricanes, extraordinary earthquakes, and volcano eruptions are constantly on the rise.

We believe that the Anunnaki planet Nibiru will soon find its place in the world agenda as Planet X, the outermost planet of the Solar System. The second item on the agenda of humanity will be the manned journey to Mars, and the colony to be established there.

Now, let us look into the works on Mars a little bit: The Mars-500 mission was a psychosocial isolation experiment conducted between 2007 and 2011 by Russia, the European Space Agency and China, in preparation for an unspecified future manned spaceflight to the planet Mars. (14) In 2012, Mars One Project —a mission to establish a permanent habitable settlement on Mars— was initiated by Bas Lansdorp from Holland, and the planned goal is announced as establishing the first colony in 2035. NASA, who announced the statement as "We will be on Mars, too!" in 2015, has given the date of the first manned flight as 2030. And in 2016, Elon Musk, the owner of the space corporation Space X, has announced his plans for both human and cargo transportation to Mars, declaring 2024 as the target date. Several countries including China and the United Arab Emirates are conducting studies on Mars. In the light of all this, we can think that a human colony will have been established on Mars before the end of the first half of 21st century arrives.

Deciphering the cuneiform, and developments about the ancient history

The change and development in Europe has manifested itself in the field of history as well. When the Italian voyager Pietro Della Valle, mentions about a weird script found in the ancient city of Persepolis, in a letter he sent from Shiraz in Iran in 1621, Europe gets introduced to cuneiform for the first time. In the years to come, this script would be deciphered, and the secret history of the ancient civilizations would be revealed. (15) The inscription Della Valle finds weird gets published in 1674 by a French man called Chardin. It gets named as "Cuneiform Writing" by Engelbert Kämpfer in 1700. (16) When the German scientist

Carsten Niebuhr, who had visited the city of Persepolis in Iran, carries the copies of some of the cuneiform inscriptions from Behistun to Europe, and publishes them in 1788, the three different types of writing on these inscriptions puzzle the scientists. (17)

After a while, the European archaeologists storm into the lands of the Ottoman Empire in the Middle East. In 1843, Paul Emile Botta, Consul of Bagdad, initiates an excavation at a location nearby Mosul known today as Khorsabad. During the excavation, he brings the Palace of Dur-Sharrukin, which carries the name of the Assyrian Emperor Sargon II, into daylight. Next to this, he discovers many tablets, inscriptions, and other antiquities. (18) Then, in 1846, the British Researcher Sir Austen Henry Layard discovers the Assyrian capital Nineveh at a location called Kuyuncik further down the Tigris River. Thousands of antiquities, monuments, statues, and cuneiform tablets collected from these ancient metropolises get carried by ships to Europe. (19)

In the works to decode the Behistun Inscription the happy ending arrives with the alphabet the Irish Orientalist Edward Hincks deciphers in 1850. In this way, the Akkadian language gets deciphered. And once these tablets taken out of Iraq were read with ease, it was discovered that there were two separate ancient states: Assur and Babylon. (20) Sir Henry Rawlinson, who gave a speech at the Royal Asiatic Society in 1853, pointed out that the names belonging to the language called as Šumer on the tablets written in the Akkad language were neither Semitic nor Indo-European; indeed, "They seemed to belong to no known group of languages, or peoples." Hereby, we witness the language of Sumer being noticed for the first time. (21) In 1857, different circles of the scientific world begin discussing whether the Akkadian cuneiform script was deciphered accurately.

In order to eliminate this ambiguity, the Royal Asiatic Society sent an unpublished copy of a newly discovered Assyrian tablet in sealed envelopes to four of the leading orientalists who were experts on the subject. These four orientalists were the top scholars in cuneiform research: Rawlinson, Talbot, Hinks ve Oppert. The society requested that each man provide an independent translation and return it in a sealed envelope. When the translations were completed and sent individually to London, a ceremony was held, and the envelopes with the translations were opened in front of a committee. When they were compared, they agreed that they had reached at similar conclusions on the main points and general sense of the text, and all the orientalist experts confirmed that the text belonged to the Assyrian King Tiglath Pileser Fe. The scientific world accepts at this point that the Akkad language, and consequently, the Assyrian and Babylonian languages were deciphered. (22)

Immediately after, the European universities founded departments, and new institutions dealing with cuneiform. The texts were scanned, and breathtaking articles were published. In this way, great advancements were achieved in the Akkad language, and the grammar of the language was recorded. In 1869, Jules Oppert suggests to the French Society of Numismatics and Archaeology to recognize the existence of a nation and a language prior to Akkad. Pointing out that the early sovereigns of Mesopotamia took the title of "King of Sumer and Akkad" in order to proclaim their legality, he proposes that this nation be called the "Sumerians", and their land as "Sumer". (23)

Discoveries revealed through dreams, sightings, and visions

When we look into the lives of the people who serve as catalysts in this fundamental change in the West, we see that important informations come through intuitions, dreams, and visions. We would like to share these situations, which we consider as the direct interferences of the Anunnaki:

- Scientific method comes through vision (1619 ACE)

On the night of November 10-11, 1619, while stationed in Neuburg, Germany, Descartes shut himself in a stove to escape the cold. While within, he had three visions about a divine spirit revealing to him a new philosophy. Upon exiting he had formulated analytical geometry and the idea of applying the mathematical method to philosophy. He concluded from these visions that everything on Earth could be analyzed with a scientific manner. Years after, this concept gets developed as "scientific method". (24)

- The French Revolution

One of the most crucial interferences on humanity's development is the French Revolution. While the normalization efforts for religion were on following the revolt against Christianity as a consequence of the Reform movements, the French Revolution breaks out, and shows humanity another path to commit to instead of religion: Nationalism. In the meantime, keep in mind that the symbol of the French Revolution is, once again, the-eye-that-sees-everything on the apex stone of the Great Pyramid —which points to Marduk/Ra—. (Figure 3)

Figure 3 – The French Revolution

- The first light bulb, and Davy's dreams

The mention of light bulb brings to our minds Edison or Tesla, but, actually, the first light bulb is invented by Sir Humphry

Davy in 1802. However, his light bulbs have a very short lifespan. In the ongoing efforts, the British Joseph Swan and American Thomas Alva Edison, invent the longer lasting light bulb, amazingly within the same years. Later on, in 1880, Edison designs the long-lasting light bulb, which lasts for forty hours. The answer to the question of how a human being came upon the idea of a light bulb, and how he invented it, once again, lies in dreams. In current times, scientific studies about Davy's dreams have been carried out to understand this matter. (25)

- The Fish Fossil appearing in a dream

The Swiss natural historian Louis Agassiz, who is considered as the founder of the natural history as a science field, is renowned with his work "Recherches sur les poison fossiles" (Research on Fish Fossils), which was published in five volumes between 1833 and 1843. Agassiz puts effort into analyzing how the anatomical structure of a sample piece fossilizes on stone so that he can publish a certain fossil fish type. After a long period, he sees the fish anatomy he is searching for in his dreams two nights in a row, but, unluckily, he cannot remember the details in a short while after he wakes up. So, Agassiz leaves a piece of paper and pen in his bed on the third night, hoping to see the same dream. The dream comes to him as he wishes, and he makes a note of the information as soon as he wakes up. Eventually, he puts them into his book. (26)

- Coincidental (accidental) discovery of photography

We do not believe in coincidences. Therefore, we consider this to be a conscious extraterrestrial interference just like Descartes' vision, and Davy's dream. One day in 1835, Louis Jacques Mande Daguerre, accidentally places an exposed silver-coated copper plate in a chemical container. In a few days, when he notices the plate again, based on the result he sees, he invents the method he would name after his own name. This invention, which he named as "Daguerrotype" becomes official in 1839 by the approval of the French Academy of Sciences. (27)

- The founding stone of a billion-dollar textile industry is provided in a dream in 1846

Canadian Elias Howe invents all parts of the sewing machine, but cannot proceed with developing the shape of the needle. One night, he sees a dream. In his dream, he runs constantly, chased by cannibals. As people throw spears at him, instead of screaming and dying, he notices the holes at the end of the spears. When he wakes up, Howe thinks of adding a hole on the needle of sewing machine. (28)

- The Evolution Theory appearing in a hallucination in 1858

Alfred Russel Wallace was trying to understand the divergence of varieties when species are separated by geographical obstacles. He had been wondering for many years how new species were formed, but he could not reach any conclusions. In 1858, he sees various illusions as a result of hallucinations caused by a tropical fever. Then, coming together with Darwin, they present their thesis, turning the scientific world upside down. (29)

- The sighting of the serpent biting its own tail, and Benzen's Discovery

The structure of a hydrocarbon compound named benzene was puzzling the chemists. As benzene was an important raw material for making synthetic paints, the problem had to be solved right away. The chemists knew that the chemical formula was $C6H6$, but they could not discover how these twelve atoms combined. The solution comes to August Kekule with a great intuition. One day in 1865, while dozing off by the fire, he sees a dream. In his dream, he watches the atoms serpentining and coiling like snakes. Kekule wakes up just as one of the snakes was biting its own tail and forming a coil, and apprehends that he has indeed found a solution for the problematic structure of the benzene molecule. The key to the solution is the ring formed by six carbon atoms each of which is linked to a hydrogen atom. (30)

- The Periodic Table appearing in a dream

Before publishing his periodic table in 1869, in which he lists the chemical elements according to their atomic and chemical characteristics, Dimitri Mendeleyev falls asleep at his desk, at a period when he was having difficulty in finding the logic behind the pattern of the organization of these elements, and he sees the order in his dream. He announces his discovery through a dream by saying "I saw in a dream a table where all the elements fell into place as required. Awakening, I immediately wrote it down on a piece of paper. Only in one place did a correction seem necessary." (31)

- Albert Einstein is trained through dreams: 1905 – Discovery of mc2

Albert Einstein tells about a dream he saw as a teenager like this: "I was sledding with my friends at night. I started to slide down the hill, but my sled started going faster and faster. I was going so fast that I realized I was approaching the speed of light. I looked up at that point, and I saw the stars. They were being refracted into colors I had never seen before. I was filled with a sense of awe. I understood in some way that I was looking at the most important meaning in my life." Later on, Einstein states that his entire scientific career was a meditation (deep contemplation) on this dream. (32)

Let us refer to the vision he saw, too: He sees a bunch of cows huddled up against an electric fence. These cows were munching on grass, their heads tied to each other by the cables on their heads. Then, as Einstein keeps watching them, a farmer suddenly switches the fence on, and Einstein watches all of the cows jump back at the same time as they got shocked. While talking to the farmer Einstein tells him how entertaining he finds the cows jumping together all at a time. But surprisingly, the farmer says that the cows had jumped one by one. This dream causes him to discover the speed of light. Furthermore, the difference in his and the farmer's perception, makes Einstein understand that time is relative. (33)

- 1913 The dream of Ramanujan, one of the most distinguished mathematicians of the world

At age 21, Srinivasa Aiyangar Ramanujan dispatches a letter with a few pages of discoveries consisting of formulas and evidence to three prominent British mathematicians. One of these was Godfrey Hardy. After contemplating on the letter, he says, "They must be true, because if they were not true, no one would have had the imagination to invent them." and after a year, he invites him to England to work on certain mathematics projects. (36) Ramanujan, who goes to England in 1913, produces about four thousand mathematical formulas, proofs, hypothesis, and equations until his death in 1920. In a thirty-two year short lifespan, he creates the belief among mathematicians as "If Ramanujan said so, it is correct." In 1976, a notebook with his notes in the last year of his life has been found. This finding has been compared to the discovery of Beethoven's 10th Symphony. A century after, these formulas are being used to understand the behaviors of black holes. (34)

The interesting part about it all is that Ramanujan says that the Hindu goddess Namagiri appears to him in his dreams, and that these formulas were shown to him by her. He recites one of his dreams like this: "While asleep, I had an unusual experience. There was a red screen formed by flowing blood, as it were. I was observing it. Suddenly a hand began to write on the screen. I became all attention. That hand wrote a number of elliptic integrals. They stuck to my mind. As soon as I woke up, I committed them to writing." (35)

Having read the hand on the screen in Ramanujan's dream, we would like to remember the incident of the hand on the wall in Book of Daniel, Chapter 5: (36)

Daniel 5:1 King Belshazzar gave a great banquet for a thousand of his nobles and drank wine with them.

Daniel 5:5 Suddenly the fingers of a human hand appeared and wrote on the plaster of the wall, near the lampstand in the royal palace. The king watched the hand as it wrote.

The difference between these two incidents is that while one happens in a dream, the other takes place in front of the eyes of thousands during a feast. But, for us, the source is the same: the Anunnaki.

- Another discovery revealed through a dream: "Bohr Atomic Model"

Niels Bohr, the Danish scientist, and the father of quantum mechanics, talks about a dream leading him to discover the structure of the atom. Bohr decides to understand the structure of atom, but cannot proceed as he wishes no matter what he tries. One night, he sees a dream while asleep about the atoms. In his dream, he sees the nucleus of the atom and the electrons turning around it just like the planets turning around the sun. As he wakes up, Bohr understands that the vision is true. He returns back to his laboratory right away, and begins searching for evidence. When he observes that it is real, he presents his discovery to the scientific world. Bohr receives the Nobel Prize in Physics, 1922. (37)

- The dream that saves Hitler during World War I.

In 1917, during World War I, while Hitler was asleep together with a battalion of tired soldiers, he sees himself covered in dust, and while being pulled into mud. Quickly, and in fear, he runs out of the trenches they were sleeping in. As he tries to recover from the dream, the building he had left a moment ago gets hit, and everyone inside dies. Had Hitler not seen this dream, the course of history would have changed completely. (38)

- The formula of insulin provided in a dream

The Canadian physicist Frederik Grant Bantin was studying diabetes. It is told that he saw a dream in one of the nights. He sees a formula on a clear surface in his dream, and he keeps it in his mind with all the details. When he wakes up, he records down the formula. The next day, when he checks it, he understands that it is the "insulin" formula. This discovery brings the Nobel Prize in Physiology or Medicine, 1923 to Dr. Bantin. (39)

- And another interference through a dream from us: "Tell Mustafa not to fear..."

During the most critical period of the Sakarya War; a time when the sounds of the cannons were even heard from Ankara, and when even the Grand Assembly of the Nation was considered to be moved to Kayseri, one morning, Ataturk asks Sergeant Ali Metin to call in Fevzi Cakmak. When Fevzi Pasha comes near Ataturk, Ataturk hands him a piece of paper and a pen, and asks him: "Write down the dream you saw today, and give it to me."

He takes another piece of paper and pen, and starts writing down the dream he himself saw that day. Then, they exchange the papers, and after reading them, they smile at each other with joy. Because both of them had seen the same dream that day. In the dream, the Prophet had told Hadji Bayram Veli: "Tell Mustafa not to fear; the victory will be theirs at the end!" As known, in those days, those two victorious commanders' names were Mustafa Kemal and Mustafa Fevzi. (40)

- Another formula appearing in a dream: The formula of Otto Loewi

In 1936, one of the German chemists, Otto Loewi was researching the effects of various chemicals on the nerves. The research had made him quite tired. One night, in his dream, he sees in all clarity the formula, which had left him sleepless for days in a row, and which he could not find no matter how hard he tried. He was even told "Here it is; the formula you are seeking!" in his dream. As soon as he wakes up, he begins writing down the formula, but, unluckily, he notices that he had already forgotten some part of it. How weird it is that he sees the same dream the next night. This time, when he wakes up, he runs to his laboratory before washing his face, and records the formula in its entirety. Thus, he makes his discovery called the "chemical transmission of nerve impulses". This brings him the Nobel Prize in Physiology or Medicine, 1936. (41)

Developments on Spiritualism

There are too many developments in this field, but we will continue with our subject after giving the two most apparent examples:

- Blavatsky founds the Theosophical Society in 1875.

The development in the West appears in the field of Spiritualism just as in any other field. Through the Theosophical Society founded by Helena Petrovna Blavatsky, the Western world gets introduced to terms like; Akasha, Akashic records, Astral body, Astral plan, Astral projection, Atman, Aura, the colors of Aura, Auric egg, Atlantis, Buddha, the Great White Brotherhood, Chakras, Healing, Dharma, Dhyan-Chohans, Thought form, Book of Dzyan, Elementals, Ether, Planetary chain, group soul, Indian Theosophy, Hyperboreans, Hierarchy, Kali-Yuga, Channeling, the Law of Karma, Karmic Plan, Karmic residue purification, causal body, causal plan, Krishna, Lemuria (The Continent of Mu), Maitreya, Mental body, Mental plan, Manu, Manvantara, Nirvana, Shamballa (Agartha), Unmanifested states, Third eye, Vision, Seven rays, Seven Root-Races, and Ascending Arc. (42)

- Right after, in 1885, Alexandre Saint-Yves d'Alveydre says that he had established telepathic contact with Agartha, and after a year, he publishes his work called "Mission de l'Inde en Europe". Europe gets introduced to Agartha in this way. (43)

- Spiritualism is defined as a field of science in National Socialist Science. For this science, during his time, Hitler says: "There is very likely a Nordic science, and a National Socialist science, which are bound to be opposed to the Jewish science. The science recognized in the West, is a spell to be broken." (Let us remember that the Nordic Extraterrestrials are the Anunnaki.) In three years, National Socialist Science gets published as three thick books, forty other simpler books designated for the public, and hundreds of brochures. And a high circulation, monthly magazine called "The Key to World Events" was published. Following World War II, a decision is taken: science should be handled single-handedly, and, thus, spiritualism is taken out of science.

Other important developments

There are many other discoveries and inventions, which we think that have happened by the interference of the Anunnaki, but for which we cannot provide any data. People get lots of information through dreams, sightings, intuitions, visions or inspiration in current times, too. Now, let us look at the developments, which we consider as interferences:

- The first automobile gets designed. The first self-moving vehicle gets produced by Ferdinand Verbiest in Beijing between the years 1679 - 1681 as a toy steam train for the Chinese Emperor. Verbiest tells how this vehicle operates in his work "Astronomia Europa", which he wrote in 1668. In 1769, French Nicolas Joseph Cugnot makes Ferdinand Verbiest's thought reality. He achieves in making the vehicle work on October 23. It operates with a steam boiler, and he calls it "fardier à vapeur" (steam freight car). (44)

- The first typewriter gets invented by William Austin Burt from Detroit in 1836. Later, Edison makes the electric typewriter in which the type bar moves with the help of an electromagnet, and he gets its patent in 1872. (45)

- The first subway opens in London in 1863. (46)

- The idea of a credit card, and the notion of using a card for purchases is defined in 1887 in the utopian novel "Looking Backward" by Edward Bellamy. (47)

- French Paul Cornu flies the first engined helicopter in 1907. (48)

- Television —the easiest way to influence masses— gets invented by John Logie Baird in 1923 in the town of Hastings in England. And the first television picture gets broadcasted in 1926 also by Baird. The quality of the pictures, which were trembling and dotted at the beginning, are developed lter by Baird. (49)

- The Charga-Plate, which was developed in 1928, is the precedent of the credit card, and it stays in use in USA starting in 1930s up until the end of 1950s. (50)

- The first photocopy machine gets invented by the American physicist Chester Carlson in 1938. (51)

- The first electronic computer ENIAC gets invented by John Mauchy and Presper Eckert in 1946. (52)

- Israel is founded in 1948, and the first Israeli-Arab War Is experienced. (53)

- The first artificial satellite of the Earth, Sputnik 1 is put into orbit by USSR on October 4, 1957.

- The internet is invented in 1958 so that the American Army can communicate between themselves.

- The Israeli-Arab Six Day War is experienced in 1967, and it ends with the victory of Israel.

- Neil Armstrong becomes the first man to land on the Moon in 1969.

- The first cellular phone is invented in 1973 by the Motorola engineers John F. Mitchell and Martin Cooper.

- Microsoft is founded by Bill Gates in 1975.

- The first Intifada, which has the characteristic of the Arab nation's revolt against the Israeli occupation, begins in 1987. The resistance, which loses ground in 1991, comes to an end in August 1993, as a consequence of the Oslo Accord signed by the Palestine Autonomy in Norway.

- A totally new era begins in 1989 upon George W. Bush's presidency in USA, and the fall of the Berlin Wall.

- Humanity gets introduced to its ancestors for the first time in 1989, and initially perceives this as a threat: the Incident of Phobos 2.

- The Gulf War begins in 1990 as a consequence of Iraq's invasion of Kuwait.

- The Warsaw Pact is ended in 1991; USSR dissolves.

- The Maastricht Agreement, otherwise known as the European Union Agreement, comes into effect in 1992.

- The first step into artificial intelligence: Chess computer Deep Blue beats the World Chess Champion Garry Kasparov in the first game in 1996.

- The information transmitted by the Galileo spacecraft in 1996 reveals that the satellites of Jupiter might have water.

- Google is founded by Larry Page and Sergey Brin in 1998.

- The second Intifada or the El Aqsa Intifada begins in 2000, and comes to an end in 2005 by the Sharm el-Sheikh Summit.

- USA starts the war in Afghanistan after September 11, 2001.

- USA invades Iraq in 2003.

- Facebook is founded by Marc Zuckenberg in 2004.

- YouTube is founded by Steve Chen, Chad Hurley and Jawed Karim in 2005.

- Saddam Hussein is executed in 2006.

- Gaddafi is lynched to death in 2011.

- The Syrian civil war starts in 2011.

- Sisi becomes the president of Egypt on May 29, 2014.

- Nuclear Accord with Iran in July of 2015

- Sisi-Trump-Selman Summit on May 20, 2017

- USA threatens Pakistan on October 5, 2017

- USA disavows the Accord on October 14, 2017

- The King's intervention in Saudi Arabia on November 5, 2017

- The Lebanese Prime Minister Hariri resigns in Saudi Arabia on November 5, 2017

- Trump announces East Jerusalem as the capital of Israel on December 6, 2017.

- The third Intifada has no returns; December 2017.

Right. So what will happen next?

This can only be answered by the planners. And, we can only guess the future by looking into the past, because the future is hidden inside the past.

Today, 90 % of the world's population is not even aware that the world is being ruled single-handedly. If we come to realize how we have become the slaves of English as a single language; of the Culture of Hollywood, and Shopping Mall Culture as a single culture; and the "moderate" world religions, in which everything is allowed.

Now, in order to be able to foresee the future, let us summarize what has happened from the past until today:

We see that almost all the wars happening within the last fifty years are in the Islamic territories. Especially after the September 11 attacks, the Islamic religion has been explicitly placed on the target board. We can say that even more difficult periods are awaiting the Muslims. The purpose is to make Islam continue its path just like the other religions; distant from radicalism, fully supporting imperialism, and as converted into a social culture. The new world order, the Greater Middle East Project, the Expanded Middle East Project, and the Moderate Islam Project are all steps towards achieving this goal.

A rapid development has been experienced within the last five hundred years. Especially within the last hundred years, this development has multiplied by four to five times, and in the last decade, humanity has been waking up to a new discovery or information daily. We do not think it would be wrong to say that more information and more technology will be waiting for us in the upcoming years. The most apparent goal of the last couple of years seems to be establishing a human colony on Mars. We can say that this goal will be reached at before the first half of the 21st century is over.

When we check into the cultural future, we can see that the world is becoming one in language, literature, poetry, cinema, and theater. Today, all the people of the world can connect, communicate, and even marry with each other as if they are the members of the same government. The cultural barriers differentiating the humans are being rapidly removed.

Let us force our imagination to foresee what might happen after 2018, the date when this book was written:

- For instance, probably we will not be surprised to hear an announcement in 2020 that cloned people live among us.

- The ninth planet might be announced as Planet X in 2023.

- Due to its elliptic orbit, Planet X might be taken out of the list of planets in 2030, just like Pluton, because, according to the new description a planet cannot cross another planet's orbit.

- We might see the first manned colony on Mars in 2032.

- 2033 might be the first year when a successful head-transplant operation is carried out. The head transplant operation of 2017 probably will end up unsuccessful, but, once on this path, for sure, one day success will arrive.

- In 2045, we might be hearing that the people, who were put to sleep, are being awakened, and we might be listening to the explanations of Walt Disney as the first person to be awakened.

- By 2050, we can already imagine the entire humanity getting introduced to the Anunnaki.

- In 2060, we might be listening to Enki's declaration to the world.

As the Age of Pisces comes to an end, and as we approach the Age of Aquarius, by looking at the developments of the last five hundred years, we can say that the new age will be the age of technology and science.

Sources:

1. Elçin, Şükrü, Halk Edebiyatına Giriş, Akçağ Yayınları, 1993, Ankara, s.72
2. http://www.greatseal.com/mottoes/coeptis.html
3. Anunnaki, insanlığın kontrolünü eline aldı? http://in5d.com/this-is-how-the-anunnaki-took-control-of-humanity/
4. https://en.wikipedia.org/wiki/Constellation
5. Heredot, Tarih, İş Bankası Yayınları S. 288
6. http://www.sozcu.com.tr/2016/yazarlar/yilmaz-ozdil/1453-1251266/
7. https://tr.wikipedia.org/wiki/Kopernik_Günmerkezliliği
8. https://tr.wikipedia.org/wiki/Galileo_Galilei
9. https://tr.wikipedia.org/wiki/Johannes_Kepler
10. https://tr.wikipedia.org/wiki/Isaac_Newton
11. http://www.history.com/this-day-in-history/william-hershel-discovers-uranus
12. https://en.wikipedia.org/wiki/Neptune
13. https://tr.wikipedia.org/wiki/Plüton
14. https://solarsystem.nasa.gov/news/2016/01/21/caltech-researchers-find-evidence-of-a-real-ninth-planet
15. http://www.esa.int/Our_Activities/Human_Spaceflight/Mars500
16. https://www.britannica.com/biography/Pietro-della-Valle
17. http://www.iranicaonline.org/articles/kaempfer-engelbert
18. https://en.wikipedia.org/wiki/Carsten_Niebuhr
19. http://www.newworldencyclopedia.org/entry/Paul-Émile_Botta
20. http://www.newworldencyclopedia.org/entry/Austen_Henry_Layard

21. https://en.wikipedia.org/wiki/Edward_Hincks
22. http://www.iranicaonline.org/articles/rawlinson-ii
23. BOTTERO, Jean, Mezopotamya, 2. Basım, Dost Kitabevi, Ankara, 2012, S.82
24. SİTCHİN, Zecharia, 12. Gezegen, Ruh ve Madde Yayınları, İstanbul, 2001
25. https://tr.wikipedia.org/wiki/René_Descartes
26. https://www.famousscientists.org/7-great-examples-of-scientific-discoveries-made-in-dreams/
27. http://www.cosmicdreaming.com/inmemoriam/elovitz.html
28. https://www.famousscientists.org/7-great-examples-of-scientific-discoveries-made-in-dreams/
29. https://en.wikipedia.org/wiki/Daguerreotype
30. http://www.cracked.com/article_20498_5-famous-things-you-wont-believe-were-invented-in-dreams.html
31. https://www.famousscientists.org/7-great-examples-of-scientific-discoveries-made-in-dreams/
32. https://www.famousscientists.org/7-great-examples-of-scientific-discoveries-made-in-dreams/
33. https://www.famousscientists.org/7-great-examples-of-scientific-discoveries-made-in-dreams/
34. https://gaiadergi.com/wp-content/uploads/2016/04/Dünyayı-değiştiren-rüyalar.docx
35. http://www.cracked.com/article_20498_5-famous-things-you-wont-believe-were-invented-in-dreams.html
36. https://eksisozluk.com/srinivasa-ramanujan--514968
37. https://tr.wikipedia.org/wiki/Srinivasa_Aiyangar_Ramanujan
38. http://www.yolgosterici.com/tevrat/tevrat28.htm

39. http://www.world-of-lucid-dreaming.com/10-dreams-that-changed-the-course-of-human-history.html
40. http://www.hurriyet.com.tr/tarihi-degistiren-ruyalar-38764431
41. http://www.ahmetturkmenoglu.com/?&Bid=853030
42. Ahmet Gürtaş, Atatürk ve Din Eğitimi, s.160-161
43. http://www.dreaminterpretation-dictionary.com/famous-dreams-otto-loewi.html
44. https://en.wikipedia.org/wiki/Helena_Blavatsky
45. https://en.wikipedia.org/wiki/Alexandre_Saint-Yves_d%27Alveydre
46. http://monumerique.aquitaine.fr/2010-2011/collegepeyrehorade/chronologie/fardieravapeurdecugnot.html
47. http://geo.msu.edu/extra/geogmich/burt.html
48. https://tr.wikipedia.org/wiki/Londra_metrosu
49. https://en.wikipedia.org/wiki/Credit_card
50. https://tr.wikipedia.org/wiki/Paul_Cornu
51. http://www.bbc.co.uk/history/historic_figures/baird_logie.shtml
52. https://en.wikipedia.org/wiki/Credit_card
53. https://en.wikipedia.org/wiki/Chester_Carlson
54. https://en.wikipedia.org/wiki/ENIAC
55. https://tr.wikipedia.org/wiki/1948_Arap-İsrail_Savaşı

SECTION 10

AS ENKI'S AGE OF AQUARIUS BEGINS

The truth can never be found by seeking, yet only seekers find it.

Bayezid-i Bistami

All that happened; was it destiny or kismet?

Since the times when the Sumerians gave writing as a present to humanity, destiny and kismet have been among the top in matters of curiosity. Although these two terms are used interchangeably nowadays, there was a distinct differentiation between the two in the texts six thousand years ago. In the Sumerian language, "Nam", meaning destiny, stands for the unchangeable course of predetermined events whereas "Namtar", which is translated as kismet, refers to the changeable course of predetermined events. The syllable "tar" means to cut, to break, to change. Sumerians believed that their kismet was created and ordered on Earth, and that it was changeable.

The Sumerian belief in destiny is adhered by many people in our current times. Based on this belief, to come to this world is our destiny, and the choice for location, gender, nationality, religion, family, and social environment is already set when we are born. Although we question everything imposed on us by the social, political, and economic systems during our childhood, as time goes by, it is suggested that we choose a path of compliance to fate, and be happy with whatever we have. To be able to have the power to change some things, we are presented with two alternatives: to study or to make money. It is as if all is connected to a mechanism, and the worldly life has an order of itself. To have a career, to marry when the time comes, to have children, a house, a car, insurance, and so on… and then; old age, and the absolute end is our destiny. Of course, there are exceptions, but we are talking about the expectations of the society here.

Although these experiences are perceived as part of a maturation period with a predetermined purpose and target, especially the choices, mistakes, and flaws of the past become the intense subject of contemplation with growing age, making one ask "I wonder if I could have changed some things?" As the answer is impossible to find, all the responsibility is put on terms like destiny, kismet, rotten luck, having one's share. When such is the case, other questions pop up in our minds: "Were the decisions we took throughout our lives our own decisions? Did we have the freedom to change the decisions we made? Or else; was our path already determined?" At this point this is our proposition: The path outlined for the awakening people is already determined, and as long as they follow their paths, their area of freedom will be less than other people. We will come back to the awakening people in short while, but what about the people detached from this process? How should their view on life be? To find the answer to this question, we will be going back about five thousand years in time; back to the Epic of Gilgamesh.

In the preface of the book we had provided Gilgamesh's prayers to God Utu (Shamash). Now, it is time and place to reiterate Shamash's reply to Gilgamesh, because it sheds light on the present day lives:

When the gods created man, death they dispensed to mankind;
Life they kept for themselves.
This is your destiny; things you do while still living.
But things you can influence and change are your kismet.
Enjoy! And take what you can from your kismet.
But you, Gilgamesh, let your belly be full.
Enjoy yourself by day and by night, be happy!
Every single day make a wedding of happiness;
Day and night, dance and play!
Let your clothes be fresh,
Bathe your head, bathe yourself in water.
Cherish the little child that holds your hand,
and make your wife happy in your embrace;
for such is the destiny of man.

Although it seems like a bitter reality that mankind's destiny has not changed since the times of the Epic of Gilgamesh, in fact, it is the most beautiful gift given to mortal man; to accept death, and to live with it; to adorn the times alive with "moments of happiness" or by turning them into quality time by "seizing the moment". Since that time mankind has kept on playing a game of happiness in an effort to put some meaning to the life s/he leads. The lesson to be learned from the Epic of Gilgamesh is that destiny weighs heavier than kismet. Based on the account; Gilgamesh's kingdom is predestined, but his escape from mortality is not.

For many generations, some have always tried to escape death, but, as far as what we are told, no mortal being has achieved this goal. It has been constantly told that humanity has no choice but to die. Furthermore, death and after-death have been portrayed to be "attractive" in my doctrines. We do not discuss the benefits of these doctrines, because the other alternative would not only cause the order to be ruined, but also create great wounds in the majority of people's lives due to the factor of "meaninglessness" of life. In fact, let us even add this at this point: there is great benefit in man's expecting and chasing something throughout his/her life time, because for the man who can reach or access everything, life would be boring and meaningless. Our greatest example for this subject is the King of Israel, Solomon. Despite being in touch with Yahweh throughout his life time, Solomon, who had the popularity of his times by having wisdom, wealth, kingdom, seven hundred wives, three hundred concubines, and thousands of slaves, became an atheist one day. (1) Although the authors of the holy books say "Although LORD, the God of Israel had appeared to him twice, and told him 'Do not worship other divinities!', Solomon diverted from the path of the LORD, and did not obey HIS command." to save face, Solomon tells about everything he had seen in the "Book of Ecclesiastes" he wrote afterwards. (2)

Various methods, rituals and beliefs to make people happy have been presented throughout history, because people need this. Even the Anunnaki, who execute the system of the world need this. This system established for the mortal man is a good system, and despite all its disadvantages, it still works like clock.

So, what is the purpose of this all? Why are seven billion people asked to continue living by "seizing the moment", and enjoying life, and by staying away from questioning while yielding to fate? Everyone might have a different answer. Let us assess it from the Anunnaki point of view.

We started our book with perhaps the most fundamental existential question of the age of science and technology: "Which one is the simulation: the human being, or the universe?" We have not been able to reach a definite answer. On the other hand, we have suggested that the Anunnaki, which is an extraterrestrial species much more advanced in nanotechnology in comparison to us, had the role of a catalyst in the process of transiting from Home Erectus to Homo Sapiens; and they had been instrumental for us to reach the current times by playing around with our genes several times.

We said that the lifespan of humans initially got lengthened, but later on, reduced to a maximum of one hundred and twenty; and that this was achieved through the genes. We explained that the genes in the superhuman "Adapa" model were blocked by 97% recently; that these genes formed the current junk genes; and that these are genes that are awakening. Well, for those people who have no awakening genes, and who do not say "Something is wrong on this Earth; what is being presented to us as information is missing some parts." allegiance to destiny has been a fabulous gift.

For the remaining minority, in the rest of the book, we tried to give all the information we could get a hand on. We exemplified the influences of the Anunnaki, who are known as the deities of the polytheistic religions, and who have human faces, on the current times. By looking at the hierarchical relations between themselves, we tried to introduce the royal family of the Anunnaki. In the latter chapters, we proposed that there have been Anunnaki bases on Mars, Phobos and the Moon for quite a long time, and that these bases are still in use at present. Furthermore, we wrote that there are Anunnaki bases such as Agharta and Shamballa on our Earth, and that Admiral Byrd might have visited the North Arianni Base, which is one of these bases.

We tried to show by examples that as the age of science and technology, the 21st century is preparing us for Enki's Age of Aquarius which will last for 2160 years. We articulated our belief in Prince Enki's shaping of the world in every field within the last five hundred years, and his leadership in directing the awakening people through his messages. We analyzed the information given to the current holy ones, elect ones and righteous ones through the ancient Book of Enoch, which we consider as one of those methods. We presented our ideas on how the foundations of the new age system were laid eight hundred years ago by Mevlana, who was privately trained by Shams, and the goal pointed out to mankind as being "one"ness.

We claimed that the Anunnaki were grouped into two clans during the Pyramid War 10500 years before today; that the struggle between these clans continued in the later periods; and that they had come to an agreement between themselves 4000 years ago. From then on, the ruling of the world would change hands between the clans consecutively, and the precession periods of the celestial system would be determining the ruling periods and orders of the clans. It was distributed among the clans based on the twelve constellations, and periods of 2160 years. Following Enlil's Celestial Bull, Marduk/Ra's Age of Ram representing the Enki clan had taken its place. Right after, Sin of the Enlil clan had become the leader of the Age of Pisces, and he has been presiding over the world leadership up until today together with the Council of Twelve. Now, the times for the Age of Aquarius was approaching, and the people of this age would become the important witnesses of this transition.

The works of change and development, which gets initiated five hundred years prior to each transition of age, have manifested themselves in the fields of science and technology prior to Enki's Age of Aquarius. The meaning of this was that the new age would be the age of information and knowing as apt for Enki. In the section where we told the alternative history of a period of six thousand years, looking into especially the last five hundred years gave us the hints for the period in front of us. Enki was now getting the World ready for the Age of Aquarius, which belonged to him, and it was times of change for us. To achieve this all, he had chosen certain families descending from his own clan. That is why leadership and civilization had shifted from the East to West.

Woman, who had stayed behind man for all times (except some exceptions), was experiencing equality in this age. Only three hundred years ago, laundry was being washed in rivers; the distance between Artvin and Istanbul was being covered in three months; people were dying of illnesses such as cold; and life was being experienced in a very simple way in many fields. However, towards the Age of the Aquarius, technology and science was presented to the service of humanity, and it had made life incredibly easier. The reason for this laid in Enki's character traits, because he was a man of science; and according to the tablets, his job had been giving information to humans at all times. Likewise, he was wishing to continue giving all kinds of information to mankind during the Age of Aquarius, which belonged only to him, because he wished for them to track the correct information by use of variables such as the mind, intelligence, logic, feelings, and inner power. In short; he had presented the meal of information. Whoever wished could take it, and begin his/her new life filled with awareness, and those who did not want it, had the total freedom to continue with their old ways of life. The awakenings of masses, which had started five hundred years ago, had changed direction by time, and they appeared as individuals' development in current times.

Sure enough, the priority for the Anunnaki at all periods was extraction of gold from the Earth, and its systematic transportation to Nibiru. To achieve this, world's economic system has been restructured during Enki's period. While the gold was being collected by giant international corporations, and sent to Nibiru, the economy was being managed over the monetary funds corresponding to the non-existent gold. And the secret organizations and the banking system played an intermediary role in this.

This must be well acknowledged: the Anunnaki have never been divine. They were only the ancestors of the mankind, and were technologically much more superior. They were simply trying to find the best system that would secure both billions of people living together, and their gold system. The system in the Age of Bull in which the Anunnaki were deemed as "Holy Masters" was the work of Enlil. The Anunnaki, who became deified after Inanna's claim to be divine, turned into the deities of the polytheistic religions upon Marduk's wish in the Age of Ram. The system in the Age of Pisces was the system, which was deemed as the best by Sin. And now, the Age Aquarius was approaching, and Prince Enki was trying to establish the best system for both the Anunnaki and the humans. Of course, while doing this, he was paying attention to not breaking the functioning system of the world, and therefore spreading the changes over a period of five hundred years, and trying to create a careful transition.

The single purpose of all the disasters, viruses, and diseases, which seem negative to us at the moment, was to keep the population of humans at a certain level, and thus, lower the risk of environmental disasters, and atmospheric problems to the minimum.

Throughout history the works on balancing the continuously increasing human population was being managed over earthquakes, diseases, and hunger. However, in this new age, Enki was trying something new: a human model which chooses between living and dying. While people, who chose to live, were finding a way to live under any circumstances, those who did not inherently enjoy living, were not even aware of the fact that they were ending their lives. Wars, hunger, diseases, and natural disasters existed in this new age, too. However, Enki was telling to the awakening ones "Let not your spirit be saddened by the negativities of the times, for the Holy One has appointed days for all things." through the Book of Enoch.

The Age of Aquarius to last 2160 years under his ruling would be simply the age of science and technology. Therefore, at the dawn of the Age of Aquarius, contrary to the Age of Pisces, religions should have been replaced by science and technology. It could be observed that the Age of Aquarius would be the age of science and technology; while steam power, machine power, electricity, telephone and so on were at full speed, it could be observed that the Age of Aquarius would be the age of science and technology. As it had happened at the turn of each new age, the Mesopotamian cities were being destroyed one after another once again. And one more time, the Golden Age was being expected. Additionally, this time, the civilizations, who were underdeveloped in science and technology, were being sacrificed for the sake of balancing the population. After 2160 years, this age would end, too, and the Age of Capricorn would begin. No one knows whether the Age of Capricorn will be an age of polytheistic religions like the Age of Ram; or monotheistic religions like the Age of Pisces; or atheist, and scientific and technological like the Age of Aquarius. But, there were two destinies known: the change would once again begin in the cities of Mesopotamia, and the expected "Golden Age" would once again not arrive. Who knows; perhaps, "The Golden Age", which found a place for itself in languages at all times, was nothing but a simple misunderstanding, and the correct version was "The Age of Gold-Collecting", which had been ongoing anyway for the last four hundred thousand years...

In this new age, one had to "seize the moment", and positive thinking had to be in every aspect of our lives, but one should never stop questioning and searching. Development for "awakening" people, who felt they were different than the other people, was only possible through awareness, and everyone had equal rights in this path. Factors such as social status, wealth, and power had no contribution to awakening. Someone with billions of dollars had to go through the same difficulties as a shepherd in the village, although this path had different tests for each. There was only one truth, but the ways of reaching at it were numerous. And when we look into the common characteristics of these ways, we observe people questioning all the time, improving themselves, adding on to their knowledge, rising up as their knowledge increases, and becoming more and more humble as they rise up. Those people, who did not discriminate against people of different religions, races, and languages; who had learned that the first condition of being a good person is to have the feeling of mercy; who treat animals well; who treat plants well, who treat children well, and most importantly, who treat themselves well, were the ones who were closest to being chosen.

It did not matter which way these people chose, because the world was becoming a beautiful place with their energy. Once they saw through their awareness what was going on, and started questioning, a more difficult period was beginning for these people: to be one of the aware ones, and to adjust to the ways of the world simultaneously.

Once this was achieved, they were provided with a guiding service through various experiences, visions, dreams and intuitions, and through little miracles, they were made to feel that they were on the right path. What is meant by right path, and right person is a person acting in accordance with his/her inner truth. When the awakening people diverted from the right path, through little warnings, they were made to intuitively become aware of this diversion. The most important tests in this process were perhaps the ones on the ego, and the attractive blessings of the worldly life. And it was right at this point where information pollution was being effective. Naturally, there were also many who were being eliminated, because, although the path chosen would be followed generally by taking support from feelings, the doors would not open up, if wisdom, intelligence, and rationing were not added on top. Proceeding only internally would not always suffice, because the desirable outcome was becoming genuine or an expert. Those who became part of a group or sect, tagging along behind someone, and resting their minds in the hands of others, were going on with their lives as people who were awakened but eliminated in the first test. Although they were not mortal people, as they had been unsuccessful in the test, they were being doomed to live by repeating themselves throughout their lives.

For those who continued their journey, the next steps meant alienation in crowds, becoming silent, watching all that happens without any reaction, and introversion. Among all the tests, the most important was freeing one's self from fear, worry, negative energy, and focusing on negativity.

The actual question we needed to ask about these phases, which we have deduced from the experiences of the awakening ones, was "What was the purpose of this awakening and process of selection? What was Enki intending for with these tests?'

When we scan the past in order to find an answer for this, we come across with the fact that there are people who were taken into the heavens, who, later on, were assigned as guides for others. In other words, when we track those thousands of people, who did not die, but vanished, disappeared, or got lost, we believe that the answer is obvious. Hereby, we are referring to a different process —beginning with Adapa, and continuing with Hanok— which thousands of people were subjected to.

If we put it into simpler terms; you know you will die, and wait for the absolute ending in your bed. However, someone, who is visible to no one but you, comes, and explains this procedure to you. In the past, in such cases, people talked about the person vanishing, flying in the air, or getting lost whereas nowadays, it might be the case that the surrounding people are loaded with the information that they are dead and buried. The funeral ceremony you participate, and the moments when you see the dead body being buried are perhaps a simulation for everybody including you. While everyone thinks that he/she is dead, he/she might be traveling in a spacecraft to his/her new life; a life to be lead in high frequency, who knows; perhaps to last for fifty thousand years, in which he/she will be guiding other people. Perhaps, the Anunnaki, who have succeeded in the transfer of consciousness, are uploading the consciousness of these people into other bodies; or perhaps, they are restructuring the DNA systems of these people, and making them younger, and eventually bringing them back to their ages of thirties. These are all guesses, but on the other hand, the fact that some people do not die, but rather simply leave, is also apparent. Ilias, and Sarikiz (Fair-Skinned Maiden), who lives on Mount Ida, from our culture are among thousands of examples in this line.

When Elijah was taken up to the heavens in a vehicle creating a whirlwind, his assistant, Elisha, and the fifty prophets had not understood what was going on, and they had kept on looking for Elijah for three days between Eriha and Teleilat Ghassul. Similarly, Sarikiz was also taken into heavens in a mysterious cloud, and her father had kept on looking for her on the skirts of Mount Ida.

We wonder if Gilgamesh, who went through so many adventures seeking immortality, but in vain, did not die either. Could he be one of these people taken into the heavens? Could this epic, which has been telling us for almost about five thousand years that mankind cannot overcome mortality, be an instrument in implementing learned helplessness to masses?

Muazzez Ilmiye Cig's book "Gilgamesh: The First Hero in History" has a part like this:

"After his death, Gilgamesh is believed to have become the judge of the underworld representing the Sun God and the Anunnaki as gods. He was prayed to, and wishes were asked from him accompanied by magic and incense. (This belief of asking for help from a famous dead person has reached the present time in the form of asking help from a Saint.) An interesting letter has been discovered in the library of a priest's house during an excavation in Sultantepe in Southeast Anatolia. This priest had lived in 7th Century BCE. In the letter, Gilgamesh asks a staggering amount of gold, silver, and precious stones from an unidentified king in order to ask his friend Enkidu to make an amulet. Without doubt, this is written as a complete joke or mockery..."

This situation, which is evaluated as a joke by the living Queen of Sumer, the most precious Ms. Muazzez Ilmiye Çığ, fits well into our description of guiding people. Perhaps Gilgamesh is not dead, and he was granted long life. Thus, he might have become one of the guiding people of the Sun God Shamash.

Taking Jesus' saying as "Many are called, but few are chosen." into account, and assuming that the called ones are the awakening ones, and the chosen ones are those who have passed all tests, we can conclude with one last question: Is it possible to be among the chosen ones?

Our views on this subject, which we have talked about in detail in our book "The Last Call: Contact with the Anunnaki" are such: while death means the absolute end for the majority of the world's population, for the little minority consisting of the awakening ones, it does not mean the end. The purpose of all the tests is to change the direction of the awakening towards exaltation, and to find the people who will be among the chosen ones. This is not an easy task, but the methods have been explicitly provided in almost every other religion and belief. The awakening people must be able to understand what the experiences they go through serve; see the signs appearing internally; and walk without fear in the path shown to them through the mind, intelligence, logic and feelings. They must be able to understand that all the blocks, and infatuating and negative situations actually serve their purpose. The awakened person, who continues on his/her path with confident steps as an individual, must also put effort in understanding the most accurate decisions for the entire humanity, the world, and the animals, by making a connection with the collective conscious. This sentence is a very beautiful saying: "Let it happen, if it serves the benefits of the whole."

Let us end our book with the sentences mentioned for Enki, the ruler of the Age of Aquarius in front of us, in the last parts of the Revelations Section of the Book of Enoch:

Who is there of all the children of men, capable of hearing the voice of the Holy One (Enki's) without emotion? Who is there capable of thinking his thoughts? (Who can have empathy with him?) Who is there capable of contemplating all the workmanship of heavens (the Solar System)? Who may comprehend them (the Anunnaki) by looking at heavens? Who is there able to see such a spirit (the Anunnaki with high frequency), describe it, ascend and watch their activities, think of them, or act like them? Who of all men is able to understand the breadth and length of the earth (the whole truth about the world)? To whom the dimensions of all things have been revealed? Or who may know the extent of heavens, its elevation, and on what it is founded on, the numbers of the stars; and where all the luminaries remain at rest?

Sources:

1. Kitabı Mukaddes I. Krallar Kitabı, 11. Bölüm 1-13
2. Kitabı Mukaddes Vaiz Kitabı

APPENDIX-1

THE ANUNNAKI IN THE ENLIL CLAN (EAST)

Inanna presenting the new King Ur-Nammu to her father Sin

THE MOON GOD SIN/ NANNAR/ EL

Sin, who is known as the Moon God, is the first Anunnaki born on Earth as the son of Enlil and Ninlil. Otherwise called as El and Nannar, Sin married to Ningal who gave birth to twin children. Inanna and Shamash are his children. In places where there is worship to the moon, Nannar/Sin is respected in the form of his dear wife NIN.GAL (Great Lady Moon Goddess). Nannar's name in the Akkadian/Semitic language is Sin. We have hints for identifying Yahweh as Sin: Part of the Sinai Peninsula being named as the Sinai Desert; Moses' first time meeting with Yahweh being here; the great manifestation taking place on Mount Sinai. In the plains right in the middle of the Sinai Peninsula, the main settlement area near the location which we think is the real Mount Sinai is still called as Nakhl in Arabic. This is the name of goddess Ningal whose name in the Semitic language is Nikal.

In the Sumerian hymns, Nannar is described as "The determiner of destinies in heavens and on Earth, the leader of all creatures alive, the one causing reality and justice take place."

The sovereignty over Ur, the best known city state of Sumer, is given to him. Another center of worship is a city we know very well: Harran. Both Sumerian texts and archaeological evidence indicate that Sin and his wife escaped to Harran, to the city of Hurri, protected by a few rivers and a mountainous region.

Ur has always remained as a city dedicated to Nanna/Sin. Harran-Urfa has been his abode for a long period. It was built on the model of Ur anyway; its temples, buildings and streets were almost identical. Andre Parrot, a French archaeologist renowned for his Mari excavations, and who had specialized in the Near East has summed these similarities in these words: "There is every evidence to support that the cult of Harran was nothing but an exact replica of that of Ur." When the Temple of Sin in Harran, which was rebuilt over and over again in a millennia, was brought to day light following the excavations lasting for fifty years, there were four stelea among the discoveries. These stelea had a very interesting inscription carved on them: The last Babylonian king Nabunaid mentions that Sin had helped him directly when he became the king. In return, he had built E.HUL.HUL ("house of great joy"), Sin's Temple in Harran, and announced Sin as the Supreme God. (1)

Nabunaid also claims on these stelea that a miracle had taken place:

A miracle "that has not happened to the Land since the days of old" had taken place:

> A deity "has come down from Heaven."
> This is the great miracle of Sin,
> That has not happened to the Land
> Since the days of old;
> That the people of the Land
> Have not seen, nor had written
> On clay tablets, to preserve forever:
> That Sin,
> Lord of all the gods and goddesses,
> Residing in Heaven,
> Has come down from Heaven.

It was then that Sin was able to grasp in his hands, "the power of the Anu-office, wield all the power of the Enlil-office, take over the power of the Ea-office - holding thus in his own hand all the Heavenly Powers." Hereby, Sin's preparations for the approaching Age of Pisces is being recounted, because each turn of age begins to manifest itself gradually about five hundred years prior to the beginning. In this way, Sin assumes the title of "Divine Crescent" and establishes his reputation as the so-called Moon God.

Sumerian texts speak in adoration of Nannar's "Boat of Heaven." (Spacecraft):

Father Nannar, Lord of Ur...

*Whose glory in the sacred Boat
of Heaven is ...*

Lord, firstborn son of Enlil.

*When in the Boat of Heaven
thou ascendeth,*

Thou art glorious.

Enlil hath adorned thy hand

With a scepter everlasting

*When over Ur in the Sacred Boat
thou mountest.*

There are worship centers for Sin in places other than Ur and Harran. Jericho, one of the important cities of the Bible, is one of them. The ruins very close to Jericho, which are known as "The City of the Moon God" had shocked the archaeologists just like the ruins of Gobeklitepe did recently, only much earlier.

WORLD'S MOST POWERFUL WOMAN: INANNA

Known as Ayizit among the Turks; Venus among the Romans; Aphrodite among the Greeks; Astarte among the Canaanites and Hebrews; Ishthar or Eshdar among the Assyrians, Babylonians, Hittites and the other ancient nations; Inanna, Innin or Ninni among the Akkadians and Sumerians, or by her other nicknames and epithets among others, Inanna was the Goddess of Warfare and the Goddess of Love of all times. Her father is the Moon God Sin, and the mother is the Moon Goddess Ningal. Inanna, who had a bright reputation among the new age Sumerian deities together with her twin brother Shamash, was the favorite of her grandfather Enlil. She is the most mentioned Anunnaki in the tablets.

Inanna is an Anunnaki who is well known in Anatolia, too. She is known as Aphrodite in the west, and as Ishtar in the east. The ancient city of Aphrodisias is her most important cult center in Anatolia. Considering Homer's Iliad as a source, we come to see that Troy is also a very important center for Aphrodite.

Although Inanna was an ordinary grandchild of Anu and Enlil, she could open up space for herself among the great deities of the Heavens and the Earth through her warrior identity and unequalled efforts especially after the loss of her fiancé Dumuzi. There are numerous inscriptions, texts, hymns, prophecies and prayers proving Inanna's physical existence.

A lengthy text called "Enmerkar and the Lord of Aratta" describes how Master Enmenkar, who was the servant of Inanna, sends messengers to Aratta, and makes it yield to Inanna. (2)

The love of Inanna and Dumuzi:

Inanna was about to choose her husband, but she just could not make a decision. She had two candidates: One was Prince Dumuzi, Enki's son and the owner of Middle Africa; and the other one was Enkimdu, who was a high rank Anunnaki officer. Her brother Shamash insists that she chooses Dumuzi, but Inanna approaches the matter with her feelings rather than her mind, and she chooses Enkimdu. Neither Shamash's pressure nor her family's views on seeing this marriage as a bridge between the Enki and Enlil Clans made Inanna change her mind. Everything Dumuzi did for Inanna were in vain, too. However, when Enkimdu gives up on his love yielding to heavy pressure on him, Inanna also gives up, and finally says yes to Dumuzi. Initially, her reply to Dumuzi has to do with reasoning, but later on she falls in love with him. (3)

According to the Sumerian mythology, the Sumerian deities Dumuzi, the younger son of Enki, and Inanna, the grandchild of Enlil, decide to get married, and they ask for permission from the elders. This mutual decision makes both the Enki and the Enlil clans happy, because this marriage could serve as a good opportunity to put an end to the tensions between these two divine clans. However, that did not turn out to be the case. Dumuzi tells his step sister Ngeshtin-ana that he wants to have a child with her, so that he could retain his right to succession, and also keep his lands in the territory of the Enki clan. (4)

Ngeshtin-ana would later come to be known as the goddess of wine, and would lend her name to a beer brand in the current times. We can qualify this kind of approaches of the great dynasties, which reached our times through the tradition of the pharaohs marrying their step sisters, as the continuation of the succession rules. While asking to have a child with his step sister, Dumuzi has taken his father and the other deities as an example. In this way; by having a child with her step sister, he would take the leading position in succession, and become the king after his father. However, at this point, we can say that the plan was counterproductive, because Dumuzi insists too much when Ngeshtin-ana does not accept, and he violates her.

As the events keep unfolding, Ngeshtin-ana tells about the situation to his oldest brother Marduk, and asks him to have her taken from there. Marduk sends ten of his best men to take his sister, and to arrest Dumuzi. Marduk actually desires Dumuzi to be arrested and put on trial. If Ngeshtin-ana marries him, the problem would be resolved automatically, yet, if she does not, he would get an exile punishment similar to that of Enlil's.

Feeling great remorse in what he did, Dumuzi constantly tries to explain his justifications to Ngeshtin-ana, and asks for her forgiveness in plentitude, but the girl ostracizes him. As midnight arrives, Dumuzi falls asleep in exhaustion from remorse and begging for forgiveness. Upon seeing a dream, he falls into a terrible fear, because in his dream he sees that he was being sentenced to death. Right as he was contemplating on this, he hears the noises of the men who arrive to arrest him, and he escapes from the house hoping to be saved. On the first opportunity he catches, he sends a message to Shamash so that he can come and rescue him. Marduk's guards catch and arrest him anyway. Nevertheless, he gets freed eventually by Shamash's assistance. After a while, Marduk's guards go after him once again. Dumuzi goes all the way to the banks of the River Nile hoping to dodge those chasing him. He chooses the location of the first waterfalls of the river, where the Aswan Dam is located nowadays, as his hideout. Although he manages to escape the chasers, he cannot escape his destiny, and death finds him on the banks of that river. This was the last day of Dumuzi's life, when he loses his balance sliding over a slippery piece of rock, falls down, and hits his head "(5)

Inanna and the Seven Gates of the Underworld

Based on the text named "Inanna's Journey to the Underworld" (6) Inanna gets on her flying bird to go to lower Abzu so that she could retrieve Dumuzi's corpse. What is being referred to as the Underworld is literally the under parts of the World, that is; around South Africa. However, through time, and mythology, it derives into a surreal meaning of netherworld and the world of the dead. Hades, who is the God of the Underworld in Greek mythology, corresponds to Enki's son Nergal, who is very well known in Sumer, and his cult area is the lower parts of Africa. On the other hand, Inanna arrives by her sister Ereshkigal (Nergal's wife) in South Africa on quite a technological journey.

After Inanna leaves the spacecraft, and while passing through the seven gates, her accoutrements and attributes were taken one by one from her. She had intended for Ereshkigal's throne, but she arrives without power. Hereby, Ereshkigal thinks that Inanna was wishing to become pregnant with the sperms to be taken from Dumuzi (just like in the example of Isis and Horus). Therefore, she does not give permission to her, and allots sixty different diseases on to her. Although the tablet mentions that Inanna was dead, we believe that this information is not accurate. Because, if the Anunnaki had a technological way of coming back from death, Dumuzi would be the first to do so. In other words, while Dumuzi could not come back from death, Inanna could.

In the rest of this text, Inanna's parents become very worried with her disappearance in Lower Abzu. Sin/Nannar explains the matter to Enlil, and Enlil sends a message to Enki. Enki learns about what had happened from his son Negal, who is Ereshkigal's husband. Then, he forms two emissaries from the clay of Abzu, who did not have any blood, and who were not affected by the death beams. He sends them down to Lower Abzu so that they could bring Inanna back; dead or alive. Ereshkigal becomes very puzzled by their appearance, and she asks: "Are you Anunnaki, or are you Earthlings?"

Namtar directs his weapons of magical power to them, but they remain unharmed. Namtar takes the emissaries by the stake where Inanna's lifeless body was hanging from. The emissaries direct the Pulser and the Emitter onto the corpse; sprinkle the Water of Life on her body; and place the Plant of Life in her mouth. Then, Inanna stirs, opens her eyes, and rises from death. As the two emissaries prepare to take Inanna to the Upper World, Inanna orders them to take the lifeless body of Dumuzi with them. While leaving through the seven gates of the Lower Abzu, her accoutrements and attributes are returned to her. She orders Dumuzi to be taken to his abode in the Black Land. The similarity of this text with the seven gates in the Sufism is yet another subject matter for a different study.

We think that hereby Inanna is fainted rather than dead, and that she wakes up through some medical intervention applied by the emissaries. Another matter is Enki forming the two emissaries from the clay of Abzu. Ereshkigal must have been seeing such beings for the first time. Otherwise, she would not have asked if they were Anunnaki or Earthlings. These beings even have names; Gala-tura and Kur-jara. In the rest of the text, these beings are defined with terms such as devils, demons, and phantoms. However, we think that they are either entities of energy or organic robots.

OTHER TEXTS ABOUT INANNA

Marduk was blamed with Dumuzi's death, and as a consequence, a power war was initiated between the Anunnaki groups lead by Marduk and Inanna. This subject has been explained in detail in the Pyramid Wars Section.

In 3760 BCE, Anu visits the Earth, and he likes his grandchild Inanna very much. Upon his return to Nibiru, he gives the E-Anna Temple in Uruk, and the spaceship designated for himself to Inanna. Highly rejoiced with this, Inanna starts dancing and chanting songs, and the praises she recites for Anu come to be sung as hymns as time passes. Sitchin considers this as a matter of love affair, but we think that the nature of the matter is quite different. Anu had waived Marduk's exile, and we think that in return for this, he gives these presents to Inanna to gain her heart, too.

Once the civilizations were started to be built as per Anu's promise as "The Earth belongs to the Earthlings." Inanna claims a right on her deceased fiancé Dumuzi's heritage. Then, she asks Enki and Enlil to grant her a land of sovereignty to belong to herself. The third region gets founded in a faraway land so that it would not be mingled ever with Sumer and Egypt, and this region gets granted to Inanna. This is the Indus Civilization.

Yet another text explains how she stole the "ME"s —also known as the tablets of civilization— from Enki through trickery to use them in the development of the Indus Civilization. (7) Inanna seduces Enki using her attraction, and by way of making him drunk, she asks to see the MEs. Inanna gets a hold of the MEs seven times after they go into a competition of drinking. Finally, when Enki falls asleep, she takes the ninety-four ME's including the lordship and the kingdom, the priesthood, the craft of the scribe, the craft of well-dressing, the craft of war, the craft of music and singing, the craft of the carpenter, the craft of the metals and precious stones, the craft of the smith, the craft of the builder, the craft of the transporter, the craft of anatomy, the craft of medical treatments, the craft of controlling the floods, the craft of astronomy, the craft of mathematics and the craft of calendars, and flees.

At this point, we can think of the movie Matrix. In the movie, our hero Neo could be loaded with any field of expertise. Upon seeing a helicopter, Neo asks for the manual for it; the manual gets loaded into his brain in the form of a software, and he gets to use the helicopter. We wonder if the Sumerians also got the pioneering jobs which required expertise and the special knowledge similarly through the "ME"s? There are different claims on the word "ME": While in this text, the MEs are considered to be gadgets holding information just as flash disks, due to the usage in the Enki-Ninmah myth (8) they are also assumed to be the information of chromosomes.

Inanna has been either the trigger-puller or the participator of almost all the wars in history. Through her role as Aphrodite in the War of Troy, Homer places her right in the middle of the war. She gets involved in such a degree that she herself goes into war and gets wounded.

The Mesopotamian kings write that generally they get signs from Inanna before going to war, and by her support they conquer places.

Dur-Sharukkin's attack on locations belonging to Marduk, and also on Babylon; Naram-Sin's occupation of Egypt —once again belonging to Marduk— and the Pyramids resembles Hitler's conquer of Europe. In the inscriptions left from the Assyrian kings, they tell how they went into wars based upon her orders and in her name; how she advised them on when to attack and when to wait; how she lead the armies sometimes; and at least once, how she granted manifestation and appeared in front of the eyes of all the army. In return for their loyalty, she has promised longevity and success to the Assyrian kings. She has assured them by saying "I shall be watching over you from the Golden Room in Heavens."

Gilgamesh, who was the ruler of Uruk in 2900s BCE, and who was partially divine (born of a human father and a goddess) tells how Inanna seduces him. But Gilgamesh knows how it would end. He asks "Which one of your lovers you loved forever?", "Which of your shepherds kept you happy all the time?" He makes a long list of Inanna's adventures of love, and denies her. As time passes by, as she reaches the higher ranks of the pantheon, and thus, as she takes more responsibility in government affairs, Inanna/Ishtar begins to display even more military qualifications, and people start portraying her more and more as a fully armed Goddess of War.

We have no way of telling whether Inanna turned into a relentless warrior after losing Dumuzi or for other reasons, but we can observe that she was never again like she used to be following this trauma. From that point on, she was simultaneously being portrayed in the tablets as a warrior, a half-naked goddess of love, and Inanna, who had sexual intercourse with many men seeking pleasure. In fact, in the Gigunu House of Pleasures, each night a man would be the guest of Inanna. And, each morning, the corpse of one of these men would be taken out. She led an unbalanced and disorderly life using her beauty and attraction in all fields. Despite her fame in love affairs, Inanna, who once became the most powerful woman of the world, never got married, and never fell in love again.

SHAMASH: THE GOD OF SUN

Shamash is born simultaneously with his twin sister Inanna from Sin and his official wife Ningal. The Sun God Shamash, who is otherwise known as Baal, Hubal and Utu, has been the commander of the Sippar Spaceport for a long time. In his inscription, Hammurabi calls god with his Akkadian name Shamash, which means "Sun" in Semitic languages. Consequently, scientists have assumed Utu/Shamash to be the Sun God of Mesopotamia.

We mostly encounter Utu/Shamash with his identity as Baal, and the name of the city of Baalbek in Lebanon means "The Valley of Baal" or "The Weeping of Baal". The city of Baalbek is the second cult center of Baal following the Great Flood. When the Greeks and the Romans conquer this city, they stop calling the city under its Hellenic name Heliopolis, which means "The City of the Sun", and start using the Semitic name "Baalbek".

The name Heliopolis suggests the belief of the Greeks which consider Baalbek to be somewhat of a "city of the god of sun", and which holds it parallel to the city in Egypt named the same. The Torah also accepts that there is a parallel city to the Heliopolis in Egypt while mentioning the Beth-Shemesh (House of Shemesh) in the north and the Beth-Shemesh in the south in Biblical terms.

Jerusalem also has to do with Shamash. Before becoming Jerusalem, it was known to be the center of "The most superior El (Sin/Nannar), The Righteous One of the Heavens and Earth". Jerusalem's oldest name is Ur-Shalem; in other words: "The City of the Completed Circle". At this point the other two cities, where the cult of Sin continued to exist for many years, come to our minds: Ur and Urfa. Scientists have put forward various theories on who Shalem could be. Just as Benjamin Mazar mentions in his article with the title "Jerusalem before David's Kingdom", some point to Enlil grandchild Shamash, while others indicate Enlil's son Ninib. Nevertheless, there is no room for discussion in any of the theories on the relation of Jerusalem's roots with the Mesopotamian pantheon.

In the Epic of Gilgamesh, we observe Shamash as one of the leading heroes. During the entire epic, we come across with the dialogues of Gilgamesh and Shamash. The fourth tablet of the epic of Gilgamesh starts with the journey to the Cedar Mountains. The results of the research show that the house of the deities on the Cedar Mountains is actually Baalbek. Let us add right at this moment: the flag of Lebanon has a cedar tree symbol even today. If we mention that Baalbek is a spaceport, the spacecrafts are called as "Shem or Mu", and that they are under the control of Shamash, then, we believe the following verse lines will be more meaningful: (9)

> O Shamash!
> My wish is to go to the Land, be my supporter!
> I want to go to the Land where the cool cedar trees are lined, be next to me!
> Allow me to set up my shems, in the place where shems are raised!

According to our theory, Shamash and his sister Inanna have been the main protectors of the Turks. Similarly, Japan's sun goddess Amaterasu is no other than Shamash's wife Aya. Besides, Shamash has been transliterated into Arabic as Shams. Shams means "the sun". (10)

TESHUP: THE HITTTITE GOD OF STORMS

Other names for the Hittite's National God Teshup, which we come across with in all parts of Anatolia, are Adad and Ishkur. Adad is a powerful god; as he is the youngest son of Enlil, he is empowered with the powers of a storm god. He was respected just as the Hurri/Hittites' Teshup, the Urartians' Teshub ("the one who blows wind"), the Amorites' Ramanu ("roaming"), the Canaanites' Ragimu ("the one who shoots the hail"), the Indo-Europeans' Buriash ("lightmaker"), and the Semites' Mcir ("the one who illuminates" the skies"). Adad/Teshup is married to Hebat. He is equally famous for his popularity and being a womanizer. Almost in every corner of the world, he has found a place for himself.

Especially his existence in South America is puzzling the mainstream science. A list of gods kept in the British Museum makes it clear that Ishkur is really the divine lord of the lands far from the lands of Sumer and Akkad. As revealed through the Sumer texts, this situation is not simply coincidental. As understood from the texts, Enlil sends his youngest son purposefully so that he can become the "Local Divinity" of the mountainous regions to the north and west of Mesopotamia.

NINURTA: THE ASSYRIANS' ARCHER GOD

Ninurta, who is also known as Ishum, Ningirsu and Asur (perhaps also as the Greek's Apollon), is a great commander and a brave soldier, but his passion for music, especially for the lyre, is also frequently mentioned in the tablets. The official god of the Assyrians, Ninurta is the son of Enlil and Ninmah. He has married to his aunt Bau, and has accepted the succession after Enlil in the Nibirun Assembly. His struggles with Enki's son Marduk, and his own brother Sin are witnessed frequently in the tablets. The double-headed eagle is the symbol of Ninurta. His cult city is Girsu. We see Ninurta's role in several points in the Anunnaki's initiation of stockbreeding. The scientists agree on the matter of agriculture starting in the Fertile Crescent, but they fall short on being able to explain why they chose to start agriculture in the mountainous regions rather than the valleys. Once again, they agree that agriculture began about twelve thousand years ago by cultivating the "wild ancestors" of wheat and barley, but they are puzzled with the genetic isomorphism of the primal grain plants. A fragmented inscription reported in a work by Samuel Noah Kramer named as "Sumerian Literature Texts belonging to Niburu" says this:

Enlil climbed up the hill, and looked above;
From there, he looked down: Everywhere was covered with water just like the sea.
He looked above once again: There was a mountain with nicely scented cedar trees.
He carried the barley up, and he planted it in the soil after terracing the mountain.
Together with the vegetables he carried up, he planted the grain seeds in the soil, by terracing the mountain.

The secret of the agricultural fields formed through terracing by the "mysterious giant hands" following the Great Flood still cannot be solved. On the other hand, this text as well as other texts tell us that Enlil and Ninurta are the architects of this work. According to us, Sacsayhuamán in the city of Cusco in Peru is one of Ninurta's (his name there is Son Viracocha) agricultural terraces. And Ollantaytambo in Peru displays the most spectacular examples of terracing.

If we knew that Machu Picchu, which seems meaningless to scientists as far as its location goes, was in fact the legendary city of Tampu-tocco, and that it was founded by the water right after the Great Flood and when the waters had not yet outgone; then we can see that the terracing was just apt for keeping the soil in place. Otherwise, we can say just like the others: "Nothing grows at the top of a mountain. It must have been built to hide from the invaders." Why would they need to build a city of such refinement in order to hide from the invaders? How and from whom the Incas got this technology is yet another mystery. In fact, if we could ask the Incas, they would have said that they had taken it from the deities.

Once Enlil and Ninurta gave the agricultural products to humanity, and the appropriate agricultural lands were discovered and rehabilitated, the genetic studies on seeds and livestock began. Enlil's two siblings; Ninmah and Enki made humanity go through evolution by the use of the domestication laboratories they founded.

The Myth of Anzu and Ninurta

Ancient texts write that Ninurta was a powerful hunter, and a warrior god famous for his military skills; and about a war he went into against Anzu. (11) The text talks about an Anunnaki commander called as "Anzu" using a vehicle referred to as the "Zu or Mu", and capturing Enlil's Destiny Tablets in Nippur through illegal ways.

Although no one knows what the Destiny Tablets are, the text gives us explanations which can serve as hints:

Rites were abandoned,
A standstill prevailed; silence reigned...
The radiance of the shelter was stripped off.

At the end of this war, Zu is defeated, and the Destiny Tablets are brought back to Enlil. Anzu, who is put on trial by the Judging Seven, is given the death sentence.

In yet another Sumerian myth called as the "Inanna and the Huluppu Tree" (12) there is mention of an extraordinary tree as the nest of both Anzu, who looks like a bird, and his "evil wife" Lilith. When the tree is cut in order to make furniture for Inanna and Shamash, Anzu flies away, and Lilith escapes to "wild places uninhabited by humans". (13) At this point, Anzu's flying away can be interpreted in many ways, but based on the text above, it seems more logical that he was executed, because we eventually see that Lilith, who is left all alone at the end, is one of the most incompatible figures in the history of the Anunnaki.

Was Lilith the very first feminist of history?

While talking about the Myth of Anzu, we need to open a parenthesis for Lilith. Over time, the plots laid, and the evil plans made by Anzu turn into a fear of a "demon which flies like a bird", and which brings suffering and the plague. This fear has been carved into the memory of humanity. Some of these demons were called as Lillu meaning "howling" and "nocturnal". Their female leader Lillitu/Lilith (female devil) was depicted as a naked and winged goddess with feet resembling to that of a bird.

The majority of the shurpu ("purification by burning") texts discovered to date are formulas of spell bounding —the pioneers of witchcraft and incantation which would last for a millennia— to be used in protection from these devilish spirits. (14) Lilith is an expert in seducing and killing men, and taking away the newborn babies off the bosoms of their mothers.

Besides, the Book of Isaiah 34:14-15 says "Wild animals will mingle with hyenas there, and wild goats will bleat at one another. Lilith will settle there, and find herself a place of rest. The owls will make nests, and lay their eggs there; they will hatch their young, and gather them under their shadows. And the buzzards will come, each one with its mate." The owl is considered holy in the myth of Lilith, and the relief depicts her with owls on each side, and desert lions behind her. She holds a Sumerian version of the Ankh, which is Egypt's symbol of immortality, in each of her hands. (15)

According to Garcia Martinez, who interpreted the Book of Isaiah based on the Dead Sea Scrolls Commentary, Lilith is a plural word, and it stands for multiple demonic entities: "And I, the Sage, declare the grandeur of his radiance in order to frighten and terrify all the spirits of the ravaging angels and the bastard spirits, demons, Lilith's, owls and jackals, and those who strike unexpectedly to lead **astray** the spirit of knowledge..."

The next source in which there is information on Lilith is the Ben Sirach Text. (16) Ben Sirach Text or Ben Sirach Alphabet is an untitled medieval text inspired from the knowledge of Sirach. This text consists of the dialogues of the Babylonian King Nebuchadnezzar and Prophet Jeremiah's son Ben Sirach, and the twenty-two stories Ben Sirach tells the king. It is quite well-known nowadays because it refers to Lilith.

Based on the story, the fame of Ben Sirach's wisdom reaches Nebuchadnezzar, and he summons him to his court. A short while after, the king's son gets sick. Nebuchadnezzar says, "Heal my son, or I will kill you!" Ben Sirach gets to work right away, and he writes an amulet. At the top of the amulet he writes the names of three angels, and depicts them with their forms and symbols including the wings, the hands and the feet. When Nebuchadnezzar looks at the amulet and asks, "Who are these?", Ben Sirach replies:

"These are the angels in charge of medicine: Snvi, Snsvi, and Smnglof. After God created Adam, who was alone, He saw that it was not good for him to be alone. So, He created a woman from the earth, so that she would a wife to him, and called her Lilith. (Book of Genesis 2:18) However, Adam and Lilith began to fight right away. He said, 'I will not lie beneath you, but only on top. For I am superior to you, you are fit only to be in the bottom position.' Lilith responded, 'We are equal to each other in as much as we were both created from the earth.', and she left Adam right on the spot. She got on God's 'Magic Mu' and escaped by flying. According to us, the term "Mu", which was translated as "Ad" hereby, represents a flying object; in other words, the current UFOs. Let us continue with the rest of the text...

"When the man complained, God sent three angels to bring Lilith back. In the meantime, God said to him: "If she agrees to come back, fine. If not, I will send one hundred of her children to die every day.' The angels left, and pursued Lilith passing through the middle of the sea in the mighty waters (an island in the Red Sea or the Elephantine Island in Egypt where the pharaoh and his soldiers got drowned while following Moses) where the Egyptians were destined to drown. Finally, they found her, and told God's Word to her, but she did not want to go back. Then, they threatened her that they would drown her, but Lilith cried out loud: 'Leave me alone!', and continued: 'I only give birth to sick infants; If the infant is male, I have dominion over him for eight days after his birth, and if female, for twenty days.'" (17)

When the angels insisted, Lilith swore to them: 'Whenever I see you or your names or your forms in an amulet, I will not give death to that baby carrying it.' Lilith had also agreed to the death of one hundred of her children every day. Rumor has it that everyday one hundred demons die, but on the other hand, babies who are adorned with amulets, which have the names or forms of the three angels, survive.

Some Jewish feminists have insisted on the reappraisal of the text claiming that the depictions of Lilith in Ben Sirach Alphabet were satirical. According to these Jewish legends dating back to the times following the Great Flood, Lilith was the first bride in the plans for Adam, and as she was rejected for the sake of Eve, it is considered that she hates men. In fact, it makes more sense that she is the offender wife of evil Anzu.

In short, although Lilith is portrayed with the evil goddess image following Anzu's execution, in fact, she is a revolutionary figure revolting against the Anunnaki. She might be still going on with the fight against the Kingdom of Nibiru together with the descendants of the lineage of the ousted king Alalu. Perhaps, this is why she has been the subject of all kinds of bad fabrication and naming.

Sources:

1. Sin Tapınağı / Şanlıurfa ve Bir Kralın Rüya Görümleri http://gokturkramu.blogspot.com.tr/2014/09/sin-tapnag-sanlurfa.html
2. Enmerkar and the lord of Aratta http://etcsl.orinst.ox.ac.uk/cgi-bin/etcsl.cgi?text=t.1.8.2.3&charenc=j#
3. Dumuzid and Enkimdu http://etcsl.orinst.ox.ac.uk/cgi-bin/etcsl.cgi?text=t.4.08.33&charenc=j#
4. Dumuzid and Jectin-ana http://etcsl.orinst.ox.ac.uk/cgi-bin/etcsl.cgi?text=t.1.4.1.1&charenc=j#
5. İnanna ve Bilulu http://etcsl.orinst.ox.ac.uk/cgi-bin/etcsl.cgi?text=t.1.4.4&charenc=j
6. Inana's descent to the nether World http://etcsl.orinst.ox.ac.uk/cgi-bin/etcsl.cgi?text=t.1.4.1&charenc=j#
7. Inanna and Enki http://etcsl.orinst.ox.ac.uk/cgi-bin/etcsl.cgi?text=t.1.3.1&charenc=j#
8. Enki and Ninmah http://etcsl.orinst.ox.ac.uk/cgi-bin/etcsl.cgi?text=t.1.1.2&charenc=j#
9. Gılgamış Aslında Nereye Gitti? http://gokturkramu.blogspot.com.tr/2015/05/glgams-aslnda-nereye-gitti-bilim-kurgu.html
10. https://tr.wikipedia.org/wiki/Şamaş
11. The Myth of Anzu http://www.gatewaystobabylon.com/myths/texts/ninurta/mythanzu.htm
12. İnanna ve Huluppu Ağacı, http://www.piney.com/BabHulTree.html
13. Zecharia Sitchin, Tanrıların ve İnsanların Savaşları, S. 134
14. Zecharia Sitchin, İlahi Karşılaşmalar, S. 307
15. Fenomen Dergisi
16. https://en.wikipedia.org/wiki/Alphabet_of_Sirach
17. https://jwa.org/media/alphabet-of-ben-sira-78-lilith

APPENDIX-2

THE ANUNNAKI OF THE ENKI CLAN (WEST)

Enki and his sons Marduk, Thoth, Gibil, Nergal, Ninagal, Dumuzi

THE LORD OF THE AGE OF THE RAM: MARDUK/AMON RA

Marduk is the son and successor of Enki and his official wife Damkina. He has worn the crown in the Age of the Ram as the leader of the Council of Twelve, the world leader, and the king of the deities. He has been granted fifty names and authorities during this ceremony. These names and symbols for him become the symbols of secret organizations and healing/energy in the years to come. (1)

Marduk had several duties on Mars before coming on to our planet. At the conclusion of these duties, he descends to Earth, and he marries with Sarpanit, who is the daughter of Prophet Enoch. Together they have three children: Asar, Satu and Nabu. Asar, in other words, Osiris as he is called in Egypt, is killed by Satu (Seth) in the fight for the throne. Asta (Isis), who is Osiris' wife, becomes pregnant with Horon (Horus) following a genetic magic of Thoth. When Horus grows up he takes vengeance from Seth for his father, and he becomes influential in Egypt's governing. On the other hand, Marduk's youngest son Nabu gets to have a voice over Babylon and Borsippa.

When Heinrich Zimmern, who transcribed and translated the Ashur texts about Marduk from the clay tablets in the Berlin Museum to writing, announced his interpretations during a conference in September 1921, he caused great chaos in the religious environments. The reason was his interpretation of this text as a god's death and resurrection, in other words, as an early Jesus story. When Stephen Langdon included the English translation of the text in his work called "Mesopotamian New Year Mystery Texts" in 1923, he had titled it "The Death and Resurrection of Bel-Marduk", and he highlighted the similarities with the New Testament tale of the death and resurrection of Jesus.

But, as the text mentions, Marduk or Bel ("The Lord") did not die; he was in fact incarcerated inside The Mountain as in a tomb; the only difference being that he was entombed alive.

We believe that Marduk (Amon Ra) is one of the Anunnaki who has taken active duty during Enki's Age of Aquarius. And this in return makes him the member of the Council of the Twelve for the new age.

In the Library of Ashurbanipal in Nineveh, a lengthy text in which Marduk had recorded his own life story has been discovered. This is what Marduk tells in the text:

"I went from where the sun rises all the way to where the sun sets. I climbed over the mountains of the Hatti land..."

THE GOD OF NUMBERS: THOTH

The Sumerian version of Thoth's name is "Ningishzidda". It means the Lord of the Tree/ The Artifact of Life. He is the protector of the divine secrets of the absolute sciences both in the Egyptian and Sumer civilizations. Thoth is thought to be born of the relationship of Enki and Ereshkigal. He is one of the three people who had an active duty in the creation of the Homo Sapiens, and in giving it the twenty-third chromosome. (The other two are Enki and Ninmah.) The secret knowledge and powers granted to Thoth are symbolized by the serpents curling on each other in this god's depictions. Everyone already knows that this is the emblem of medicine and healing still in use in the current times, representing the double helix DNA. Thoth corresponds to Hermes in the Greek language.

The imagery depictions which has survived from ancient Egypt point out that Ptah/Enki's son Thoth was aware of all kinds of biological and genetic processes, and that he applies these onto his genetic skills. A mural, which depicts scenes of Pharaoh Seti I playing the role of Osiris, discovered in Abydos, shows Thoth while he is giving the dead god his life back through the Ankh Symbol, and while taking two separate DNA strands from him.

In another depiction in the Book of the Dead, which shows Horus' birth after this incident, we see two separate goddesses of birth, who are helping Thoth, while holding a strand of DNA each. The double helix of the DNA is separated from each other; while only one strand is holding the newly-born Horus, the one being shown is connected to that of Isis'.

We think that Quetzalcoatl, the great god of the Central American nations, is actually Thoth. In his descriptions Quetzalcoatl is mentioned as the Feathered Serpent or the Winged Serpent. He is also a god who knows well, and teaches the secrets of construction of temples, the numbers, astronomy and calendars. According to Sitchin, Thoth goes to Central America together with his followers in 3113 BCE, and this emigration is considered as the beginning of the Mayan Calendar. The most interesting part of this calendar is the name of the six month: "Teoleco" (based on the Gregorian calendar; the period between September 10 and 29). In the Aztec language it means "the return of the gods".

Just as the Americans make Thoth as one of theirs as "Quetzalcoatl", the Greek adopt him as "Hermes", and call him Hermes Trismegistus, which means "three times the greatest". In the meantime, there are people who say that Thoth is the same person as Idris who is mentioned in the Qur'an. This is not true. Thoth is a deity; in other words, an Anunnaki, whereas Idris is human. For those who wonder who Idris is, let us tell: he is the figure known to be Hanok or Enoch.

Thoth is the architect of all stone circles, the mathematical and astronomical calculations of the ziggurats, and the ancient buildings in the form of a circle. However, his most important structures are the three pyramids and the Sphinx in Egypt. We believe that one of the most active Anunnaki of the Age of Aquarius will be Thoth.

THE PRINCESS OF THE ENKI CLAN: ISIS

We think that Isis, whose name in Sumerian is Asta, corresponds to Artemis/Hecate in Greek Mythology. Her father is Shamgaz, one of the Anunnaki on Mars, who is an Igigi mentioned as the leader of the fallen angels in the Book of Enoch; and her sister is Nephthys, who is the wife of Seth. She becomes pregnant to Horus with Osiris' sperms by the help of Thoth when her husband Asar (Osiris) is killed by Seth. After taking over all the authorities of Osiris, she becomes the most powerful woman of the Enki clan in line with her level of ambition. Isis, whom we think will be in the Council of Twelve in the approaching Age of Aquarius, has great activeness in the world affairs today. She has a share especially in the formation of the nation of Is-ra-el together with the other two Anunnaki: El and Ra. Furthermore, Isis, whom we think was also active in the foundation of Christianity, is being compared to Mary even today.

The most important cult center of Isis is Aswan and the near surrounding which is also called as Upper Egypt. We think that the Isis Temple on the Island of Philae is her center, and the ancient cities of Ephesus, Phaselis and Lagina in present day Anatolia are her cities. Cybele and Diana are her other names.

The Temple of Artemis, one of the seven wonders of the ancient world, which is located in Ephesus, was built in her name. During the times of Theseus, who was the leading hero of Ionia, the "Amazons" would stop by here on their way to Attika to offer their sacrifices to Diana, in other words, to Artemis in the Temple of Artemis. Although there are allegations that the "Cult of Diana" existing there had been carried over by the Amazons, the public has generally adopted the view that it was transferred from the planet of Jupiter. Although the Temple of Artemis was rebuilt seven times, the cult on display at the most sacred part of the temple has always been worshipped.

The existence of the goddess, her being real and the acceptance of her existence by the believers is best proved by the fact that people have been seeing her in their dreams in all times. Strabon writes that the goddess appears to Lady Aristarche, who is a high level noblewoman from the city of Marseille, in her dream. In the dream, Artemis directs the lady for camaraderie for the Hellenistic adventurers. In another incident, Metagenes, who was one of the architects of the temple, finds a technique to elevate large stones to a certain height, but he cannot manage to place a large piece of marble above the entrance gate. When he falls asleep totally in despair, tiredness and worry due to this failure, Goddess Artemis appears to him in his dream telling him to be at ease and not to worry; and she sanctifies him. When he wakes up the next morning, Metagenes gets astonished upon seeing that the piece of marble is in place as he wished.

This situation brings to mind the kings put to duty in dreams by the gods so that they can build temples in Sumer, Akkad, Babylon and Assyria. In one such dream, God Ninurta appears in King Gudea's dreams every night, and gives him all the plans of the temple. Furthermore, he follows up in person on the construction of the temple. At this point, the visions mentioned in the Torah, in which Yahweh tells about the new temple in meticulous detail to Prophet Ezekiel, and makes him take notes about the ell calculation, make more sense. (3)

Goddess Artemis is important for the region's people to such an extent that they have dedicated a whole month of worshipping for her. This month, which we call as April, has been named as Artemision, and it has been sanctified through worship and games. And, then, at the end of the month, the Festivities of Artemis (Festivities of Diana) has been added and celebrated. Later on, the Ephesians start building this huge temple by taking help from their goddess. We will not go into the matter of how the temple was built, how the marbles were excavated, processed and put in place. For those wishing for further detailed information, we can suggest E.J. Davis' book "Anatolia", which we have consented as a source frequently. When this huge temple was finally finished, it was described as stunning. The building was extraordinarily big and spectacularly ornate. Ancient sources say "It was as if the work of entities superior to humans.", and add "From dawn to sunset, even the sun has not ever seen anything more breathtaking than this temple."

Antipater of Sidon, who compiled the list for the seven wonders of the ancient world, tells how the Artemis Temple is much more spectacular than even the Egyptian Pyramids in this way:

"I have set eyes on the wall of lofty Babylon on which is a road for chariots, and the statue of Zeus by the Alpheus, and the hanging gardens, and the colossus of the Sun, and the huge labor of the high pyramids, and the vast tomb of Mausolus; but when I saw the house of Artemis that mounted to the clouds, those other marvels lost their brilliancy, and I said, "Lo, apart from Olympus, the Sun never looked on aught so grand." (Antipater, The Greek Anthology [IX.58])

Philon of Byzance, who was awed by the height and craftsmanship, writes this for the temple:

"I have seen the mighty craftsmanship of the ancient Babylonians, and the tomb of Mausolus. But when I saw the temple at Ephesus rising to the clouds, all these other wonders were put in the shade."

Clement of Alexandria, who lived towards the end of the second century, mentions about a female oracle. According to the prophecy, "The ground will open up and shake, and the Temple of Artemis will be devoured into hell like a boat in a storm. And the Ephesians will search for the temple by the river in mourning and crying, and they shall no longer live there."

Nowadays, there is hardly anything left from this temple, which was demolished during the campaigns of the Roman Empire, which were started with the purpose of establishing the monotheistic belief. Despite the huge sizes of the stones, and the hard-to-explain amplitude of the ruins, the temple has been completed vanished; how and where it has gone is unknown. If you ever visit Ephesus, you can see that the prophecy was fulfilled to its full extent simply by looking at the present condition of the valley, and you can see the demolished city with your bare eyes. Goddess Artemis has abandoned the city, and there is nothing but a few columns left from the temple.

The earliest written information describing Isis (Hecate) is also found in Hesiodos' writings. Hecate is a Carian goddess whose name is not counted among the Greek deities. In his book "Metamorphoses" the Roman philosopher Lucius, talks about Hecate like this:

"I, mother of all Nature and mistress of the elements, first-born of the ages and greatest of powers divine, queen of the dead, and queen of the immortals, all gods and goddesses in a single form; who with a gesture commands heaven's glittering summit, the wholesome ocean breezes, the underworld's mournful silence; whose sole divinity is worshipped in differing forms, with varying rites, under many names, by all the world. There, at Pessinus, the Phrygians, first-born of men, call me Cybele, Mother of the Gods; in Attica, a people sprung from their own soil name me Cecropian Minerva; in sea-girt Cyprus I am Paphian Venus; Dictynna-Diana to the Cretan archers; Stygian Proserpine to the three-tongued Sicilians; at Eleusis, ancient Ceres; Juno to some, to others Bellona, Hecate, Rhamnusia; while the races of both Ethiopias, first to be lit at dawn by the risen Sun's divine rays, and the Egyptians too, deep in arcane lore, worship me with my own rites, and call me by my true name, royal Isis." (4)

THE FATHERLESS CHILD: HORUS

Horus has been mentioned as the fatherless child Horon in the Sumerian language. He is born as the son of Osiris and Isis.

After he takes revenge for his father's killing by Seth, Ra appoints him as the ruler of Egypt. It is thought that the corresponding figure in Greek mythology is Apollon, but there are no clear views on this subject. His mother hides Horus after his birth, and later on he gets entrusted to Gibil. While being raised by Gibil, Horus escapes death after a scorpion bite through Thoth's interventions.

THE SOUTH AFRICAN GOD: NERGAL

Nergal, or otherwise called Erra, is the same as what the Greeks call as Hades. Nergal, who is one of the sons of Enki, lives together with his wife Ereshkigal in South Africa, which is also called as the "Underworld". Therefore, the city of "Meslam" has been his cult center. By time, the term "Underworld" derives into the concept of the netherworld and the world of the dead. Nergal has been associated with Mars during the Age of the Bull. However, as he is expelled from the Council of Twelve during the Age of the Ram, he is not associated with any planet.

In a text called as the Erra Myth (5), which is thought to have been dictated directly by an Anunnaki, the events that cause Sumer to disappear are told as Nergal had them recorded. The Epic of Erra recounts the pressure on Marduk during the last periods of Sumer, and the demands from him to give up his wishes on being the leading god, from the point of Nergal and Ishum (Ninurta). From this text, we learn about the peaceful meetings with Marduk, and the discussions of the Anunnaki Assembly. Upon seeing that there is no other way to stop Marduk and Nabu, the Great Deities Assembly gives the authority of bringing out the "Seven Fierce Weapons" kept under lock in Abzu to Nergal and Ninurta. What is in mention hereby is a nuclear weapon, and not only it vaporizes the spaceport, but also creates an enormous crater on the Sinai Peninsula, and a blackened area around it. The radioactive cloud created as a result of firing the nuclear arms by Nergal and Ninurta brings all Sumerian cities to an end in 2023 BCE.

The "five sinful cities of the Torah" that were located in the Siddim Valley in the south of the Dead Sea, and which has supported Nabu have been erased from the face of this earth with this nuclear attack. The actual reason for Sodom, Gomorrah and the other cities to be demolished is their support to Marduk and Nabu.

Ninurta and Nergal, who had been in opposing forces during the Pyramid Wars in 8760 BCE, start up a very famous friendship later on. In fact, while telling about his various wars, the great Assyrian King Shalmaneser suggests that he owes his victories to the weapons provided by these two gods: "I fought wars with the "Great Might" my god Ashur gave me, and the weapons my leader Nergal has presented to me." Ashur's weapon is described as having a "frightening brilliance". During a war fought with the Adini, the enemy "sees the terrifying brilliance of Ashur, and this dumbfounds them", and they flee.

Besides, Nergal is also known as the god who watched over the south during the Great Flood. The Epic of Gilgamesh is very clear on the direction where the storm comes from: The storm comes from the south. Actually, prior to the tidal wave, which overthrows "Nergal's fore stations" in the Underworld, first comes the clouds, the winds, the rain and the darkness.

We come across with Nergal also on a big wall in Yazilikaya, Hattusha depicted as a god.

THE MINING GOD: GIBIL/GEB

The Lord of Abzu, Enki has another son, who is responsible for the fire and melting the minerals. As the blacksmith of the world, GI.BIL ("the one burning the soil") has been generally portrayed as a young god emitting red-hot beams or sparks of fire from his shoulders who is about to come out of the earth, or go into the earth. The texts mention that Gibil is equipped with "wisdom" by Enki. We see Gibil in the texts while raising Horus, and helping Marduk during the Pyramid War.

Sources:

1. Enuma Eliş Metni 6-7
   http://www.piney.com/EnumaSpeis6a7.html
2. Bel Marduk Miti ve İsa,
   http://www.bobkwebsite.com/belmythvjesusmyth.html
3. http://www.yolgosterici.com/tevrat/tevrat27.htm 40-48. Bölümler
4. http://www.mugladevrim.com.tr/kose-yazarlari/ibrahim-ergin201747123515/lagina-da-Hecate-tapinagi--osman-hamdi-bey
5. Erra Şiiri, https://therealsamizdat.com/category/poem-of-erra/

Egypt Abydos Osiris Temple

Egypt Abydos Osiris Temple

Egypt Edfu Horus Temple

Egypt Danderah Hathor Temple

Göbeklitepe - Turkey

Mount Ida - Turkey

Mount Ida - Turkey

Nemrut Mount - Adıyaman

Nemrut Mount - Adıyaman

Egypt Abydos Osiris Temple

Bibliography

1. ARİSTOTELES, Atinalıların Devleti, 3. Baskı, Türkiye İş Bankası Yayınları, İstanbul, 2016
2. BALKAN, Kemal, Babilde Feodalizm Araştırmaları Kas'lar Devrinde Babil, Ankara, Dil ve Tarih - Coğrafya Fakültesi Sumeroloji Bölümü, Tezi veren: Ord. Prof. Dr. Benno Landsberger.
3. BERLİTZ, Charles, Atlantis'in Esrarı, Milliyet Yayınları, İstanbul, 1976
4. BLAVATSKY, H.P, Peçesiz İsis 1, Mitra Yayınları, İstanbul, 2016
5. BOTTERO, Jean, Kültürümüzün Şafağı Babil, 5. Baskı, Yapı Kredi Yayınları, İstanbul, 2015
6. BOTTERO, Jean, Mezopotamya, 2. Baskı, Dost Yayınevi, Ankara, 2012
7. BOTTERO, Jean-KRAMER S. N, Mezopotamya Mitolojisi, Türkiye İş Bankası Yayınları, İstanbul, 2017
8. BOTTERO, Jean-STEVE, M. J, Evvel Zaman İçinde Meopotamya, 7. Baskı, Yapı Kredi Yayınları, İstanbul, 2016
9. BULFINCH, Thomas, Klasik Yunan ve Roma Mitolojisi, 3. Baskı, İnkılap Yayınevi, İstanbul, 2014
10. CHURCHWARD, James, Kayıp Kıta Mu'nun Çocukları, Omega Yayınevi, İstanbul, 2010
11. CHURCHWARD, James, Kayıp Kıta Mu'nun Kutsal Sembolleri, Omega Yayınevi, İstanbul, 2015
12. CHURCHWARD, James, Mu'nun Kozmik Güçleri 1, Omega Yayınevi, İstanbul, 2017
13. CHURCHWARD, James, Mu'nun Kozmik Güçleri 2, Omega Yayınevi, İstanbul, 2014
14. ÇIĞ, Muazzez İlmiye, Kur'an, İncil ve Tevrat'ın Sümer'deki Kökeni, 9. Basım, Kaynak Yayınları, İstanbul, 2005.
15. ÇIĞ, Muazzez İlmiye, Gılgameş Tarihte İlk Kahraman, 5. Basım, Kaynak Yayınları, İstanbul, 2012.

16. ÇIĞ, Muazzez İlmiye, Sümerlerde Tufan Tufan'da Türkler, 10. Basım, Kaynak Yayınları, İstanbul, 2016.
17. ÇIĞ, Muazzez İlmiye, İbrahim Peygamber, 18. Basım, Kaynak Yayınları, İstanbul, 2015.
18. DANIKEN, Eric Von, Sfenks'in Gözleri, İnkılap Yayıncılık, İstanbul, 1990
19. DAVIS, E.J, Anadolu, Arkeoloji ve Sanat Yayınları, İstanbul, 2006
20. DEMİR, Seyfullah, Tanrısal Bilim, Mavi Kalem Yayınevi, İstanbul, 2013
21. DİAMOND, Jared, Tüfek, Mikrop ve Çelik, 17. Basım, Tübitak Popüler Bilim Kitapları, Ankara, 2006
22. DROYSEN, Büyük İskender Tarihi, Dharma Yayınevi, İstanbul, 2007
23. EURİPİDES, Bakkhalar, 3. Basım, Türkiye İş Bankası Yayınları, İstanbul, 2014
24. ELDEM, Burak, 2012: Marduk'la Randevu, 7. Baskı, İnkılap Yayınevi, İstanbul, 2003
25. ELIADE, Mircea, Dinsel İnançlar ve Düşünceler Tarihi, Kabalcı Yayınevi, İstanbul, 2003
26. ERGİN, Muharrem, Orhun Abideleri, 24. Baskı, Boğaziçi Yayınları, Eylül, 1999
27. ERKUT, Sedat, Hititçe Çivi Yazılı Bogazköy Belgelerinde Geçen Sumerce Lusımug =Metal İşleyicileri hakkında, A.Ü. Dil ve Tarih-Coğrafya Fakültesi Tarih Bölümü Doktora Tezi
28. FURLONG, David, Piramitler Gerçeği, İzdüşüm Yayıncılık, İstanbul, 2008
29. GÖK TÜRK, Amon Ra-Uzaylı Prensin Yaşam Öyküsü, Mavi Kalem Yayınevi, İstanbul, 2014
30. GÖK TÜRK, Son Çağrı Anunnakilerle Temas, Mavi Kalem Yayınevi, İstanbul, 2016
31. GÖZAÇAN, Levent, Kaderini Çiz, Mavi Kalem Yayınevi, İstanbul, 2017
32. GRİMAL, Pierre, Yunan Mitolojisi, 3. Baskı, Dost Yayınevi, Ankara, 2012

33. GÜLEN, Nurdoğan K, Şuppilulima, Alfa Yayınları, İstanbul, 2010
34. HARARI, Y. Noah, Hayvanlardan Tanrılara Sapiens, 14. Baskı, Kolektif Yayınevi, İstanbul, 2016
35. HARBOTTLE, Rod, Kutsal Kitap Yerler Sözlüğü, 2005
36. HART, Will, Galaktik Gen, Sınır Ötesi Yayınları, İstanbul, 2005.
37. HAWKING, Stephen, Ceviz Kabuğundaki Evren, Alfa Kitabevi, İstanbul, 2002
38. HEREDOTOS, Tarih, 11. Basım, Türkiye İş Bankası Yayınları, İstanbul, 2015
39. HOMEROS, İlyada, 3. Basım, Türkiye İş Bankası Yayınları, İstanbul, 2016
40. İNAN, Afet, Eski Mısır Tarihi ve Medeniyeti, 2. Baskı, Türk Tarih Kurumu Basımevi, Ankara, 1987
41. KİNDER, Hermann-HİLGEMANN, Werner, Dünya Tarihi Atlası, ODTÜ Yayıncılık, Ankara, 2006
42. KESKİN, Günyüz(Çevirmen), Enoch'un Kitabı, 4. Baskı, Hermes Yayınevi, İstanbul, 2014
43. KNIGHT, Christopher-BUT
44. LER, Alan, Ay'ı Kimler Yaptı, Sınırötesi Yayınları, İstanbul, 2010
45. KRAMER, Samuel Noah, Tarih Sümer'de Başlar, Kabalcı Yayınları, İstanbul, 1999
46. KRAMER, Samuel Noah, Sümerler, Kabalcı Yayınları, İstanbul, 2002
47. KRAMER, Samuel Noah, Sümer Mitolojisi, Kabalcı Yayınları, İstanbul, 2014
48. KRAMER, Samuel Noah, Sümerlerin Kurnaz Tanrısı Enki, Kabalcı Yayınları, İstanbul, 2000
49. KSENOPHON, Anabasis Onbinlerin Dönüşü, Türkiye İş Bankası Yayınları, İstanbul, 2015
50. LEONARD, George, Somebody Else Is On The Moon, Pocket Book, New York, 1976
51. MACQUEEN, J.G, Hititler ve Hitit Çağında Anadolu, 4. Baskı, Arkadaş Yayınevi, Ankara, 2015

52. MADEN, Sait(Çeviren), Gılgamış Destanı, Türkiye İş Bankası Yayınları, İstanbul, 2016
53. MADRİGAL, Marc, Mısır'dan Çıkış ve Arkeoloji, Kutsal Kitap ve Arkeoloji Yayınları, İstanbul, 2014
54. MADRİGAL, Marc, Tevrat Döneminde Günlük Hayat, Kutsal Kitap ve Arkeoloji Yayınları, İstanbul, 2014
55. MELCHİZEDEK, Drunvalo, Yaşam Çiçeğinin Unutulmuş Sırrı 1, Butik Yayıncılık, İstanbul, 2010
56. MELCHİZEDEK, Drunvalo, Yaşam Çiçeğinin Unutulmuş Sırrı 2, Butik Yayıncılık, İstanbul, 2010
57. NEWTON, İsaac, Kutsal Kitabın Yorumu, 2. Baskı, MEDAM Yayınları, İstanbul, 2012
58. OATES, Joan, Babil, Arkadaş Yayınevi, Ankara, 2004
59. PLUTARKHOS, İskender-Sezar Paralel Hayatlar, Türkiye İş Bankası Yayınları, İstanbul, 2015
60. SEÇKİN, Renan, Astral Kapılar, 2. Basım, Mavi Kalem Yayınevi, İstanbul, 2017
61. SEÇKİN, Renan, Kahin Vanga, 2. Basım, Mavi Kalem Yayınevi, İstanbul, 2016
62. SCOGNAMILLO, Giovanni, Dünyamızın Gizli Sahipleri, 22. Baskı, Kamer Yayınları, İstanbul, 1996
63. SCHMİDT, Klaus, Taş Çağı Avcılarının Gizemli Kutsal Alanı Göbekli Tepe, Arkeoloji ve Sanat Yayınları, İstanbul, 2007
64. SİTCHİN, Zecharia, 12. Gezegen, Ruh ve Madde Yayınları, İstanbul, 2001
65. SİTCHİN, Zecharia, Kozmik Tohum, Ruh ve Madde Yayınları, İstanbul, 2000
66. SİTCHİN, Zecharia, Gökyüzüne Merdiven, Ruh ve Madde Yayınları, İstanbul, 2002
67. SİTCHİN, Zecharia, Kayıp Diyarlar, Ruh ve Madde Yayınları, İstanbul, 2005
68. SİTCHİN, Zecharia, Tanrıların ve İnsanların Savaşları, Ruh ve Madde Yayınları, İstanbul, 2005
69. SİTCHİN, Zecharia, Kozmik Şifre, Ruh ve Madde Yayınları, İstanbul, 2006
70. SİTCHİN, Zecharia, Zaman Başlarken, Ruh ve Madde Yayınları, İstanbul, 2006

71. SİTCHİN, Zecharia, İlahi Karşılaşmalar, Ruh ve Madde Yayınları, İstanbul, 2007
72. SİTCHİN, Zecharia, Enki'nin Kayıp Kitabı, Ruh ve Madde Yayınları, İstanbul, 2008
73. SİTCHİN, Zecharia, Dünya Tarihçesi Keşif Seferleri, Ruh ve Madde Yayınları, İstanbul, 2009
74. STRABON, Geographica, 8. Baskı, Arkeoloji ve Sanat Yayınları, İstanbul, 2015
75. TOSUN, Mebrure, Sümer-Babil Tanrı Sembollerinin Adları Üzerinde Bir Araştırma, Ankara Üniversitesi Dil ve Tarih - Coğrafya Fakültesi Sümeroloji Kürsüsü, Ankara, 1962
76. TOSUN, Mebrure Osman, Sümer - Babil Sanatının Bazı Önemli Mitolojik Tipleri Üzerinde Yeni Arkeolojik ve Filolojik Araştırmalar, Ankara Dil ve Tarih - Coğrafya Fakültesi Sümer Dil ve Edebiyatı Bölümü Tezi veren: Prof. B. Landsberger
77. TÜFEKÇİ, Çiçek Sekban, Ari, Destek Yayınları, İstanbul, 2017
78. TÜRKEŞ, Umay, Türklerin Tarihi, 2. Baskı, Akçağ Yayınları, , Ankara, 2007
79. VANCE, Ashlee, Elon Musk, 4. Baskı, Buzdağı Yayınları, İstanbul, 2017
80. WILLAMS, Friedrich, Ege Medeniyetleri Tarihi, Düşünen Adam Yayınları, İstanbul, 1993
81. WILKINSON, Kathryn, Semboller ve İşaretler, 2. Basım, Alfa Yayınları, İstanbul, 2011
82.

Printed in Great Britain
by Amazon